This View of Life

THIS VIEW OF LIFE

Completing the Darwinian Revolution

David Sloan Wilson

Pantheon Books, New York

Pantheon Books and colophon are registered
trademarks of Penguin Random House LLC.

Library of Congress Cataloging-in-Publication Data
Name: Wilson, David Sloan, author.
Title: This view of life : completing the Darwinian revolution /
David Sloan Wilson.
Description: First edition. New York : Pantheon Books, 2019. Includes
bibliographical references and index.
Identifiers: LCCN 2018031977. ISBN 9781101870204 (hardcover : alk. paper).
ISBN 9781101870211 (ebook).
Subjects: LCSH: Social Darwinism. Social evolution.
Classification: LCC HM631 .W55 2019 | DDC 303.4—dc23 | LC record available
at lccn.loc.gov/2018031977

www.pantheonbooks.com

Jacket design by Kelly Blair

Printed in the United States of America

First Edition
2 4 6 8 9 7 5 3 1

To all who are reaching for an ethics for the whole world

There is grandeur in this view of life, with its several powers, having been originally breathed by the Creator into a few forms or into one; and that, whilst this planet has gone cycling on according to the fixed law of gravity, from so simple a beginning endless forms most beautiful and most wonderful have been, and are being, evolved.

—CHARLES DARWIN, *On the Origin of Species*

The evil in the world comes almost always from ignorance, and goodwill can cause as much damage as ill-will if it is not enlightened . . . There is no true goodness or fine love without the greatest possible degree of clear-sightedness.

—ALBERT CAMUS, *The Plague*

Contents

Prologue

Pierre Teilhard de Chardin (1881–1955), a scientist and Jesuit priest, made an observation about humankind that departed from both Christian doctrine and the scientific wisdom of his day: In some respects we are just another species, a member of the Great Ape family, but in other respects we are a new evolutionary process. That makes the origin of our species as significant, in its own way, as the origin of life.

Teilhard lived at a time when the Catholic Church still regarded science as a legitimate path to God. Like his predecessor, Galileo Galilei (1564–1642), he was both authorized to study the natural world and restricted when his inquiry began to threaten religious dogma. Whereas Galileo gazed at the heavens, Teilhard dug in the earth for fossils. He was a paleontologist and part of the team that discovered Peking Man (now classified as *Homo erectus*), one of the first fossil skulls that provided a "missing link" between us and our more apelike ancestors. Whereas Galileo was threatened with torture, Teilhard was blocked from accepting prestigious academic positions and forbidden to publish his work. His best-known book, *The Phenomenon of Man*, was written in the 1930s and Teilhard waited patiently until the end of his life before privately arranging for it to be published posthumously.[1]

In this panoramic work, he describes earth before the origin of life as just another barren planet shaped by physical processes. Then, living organisms began to spread over its surface like a kind of skin. They too were "just" a physical process. Teilhard resisted the temptation to invoke a divine spark to explain the origin of life. But living organisms differ from non-living physical processes in their capacity to replicate and diversify into "endless forms most

Pierre Teilhard de
Chardin: Scientist and
Priest

beautiful," as Darwin wrote in the final passage of *On the Origin of Species*. Over immense periods of time, living processes began to rival non-living physical processes in shaping the planet and atmosphere, giving earth the unique appearance of a multicolored jewel when viewed from space. Using a word coined by the geologist Eduard Suess in 1875, Teilhard called the influence of life on the planet earth the biosphere.

Next, Teilhard asks us to imagine one species, located on one twig of the tree of life, that begins to proliferate and diversify much more rapidly than the other twigs. In an astonishingly short period of time, a new skin spreads over the earth, rivaling other living processes and non-living physical processes in shaping the planet and atmosphere. That species was *Homo sapiens*, and Teilhard coined a new word, the noosphere, to describe the second skin.

This word is derived from the Greek word for "mind" (*nous*), signifying that the new skin has a mental dimension in addition to a physical dimension. Teilhard described consciousness as a process of evolution reflecting upon itself. He portrayed the human colonization of the planet as starting with "tiny grains of thought" that would eventually merge with each other to form a global conscious-ness and self-regulating superorganism called the Omega Point.

The idea that we are part of something larger than ourselves, which could be considered an organism in its own right, has been expressed a thousand different ways in cultures around the world. So too has the idea that the "something" is expanding—or at least needs to—ultimately embracing all of humankind and the planet earth. Sometimes these ideas are given a religious and spiritual for-mulation, but they also lie behind practical efforts to create good governments and economies and to expand their scales. From a purely secular perspective, Teilhard's Omega Point corresponds to the vision of a world where governments work together for the good of their citizens and live in balance with the rest of life on earth. We seldom associate politics and economics with religion and spirituality, and in many ways we feel the need to keep them apart, as with the separation of church and state. Nevertheless, words such as "corporation" (derived from the Latin for "body") and phrases such as "body politic" signify that whatever we mean by the word "organism" can be applied to entities that are larger than organisms, such as a human society or a biological ecosystem.

Teilhard's cosmology was new, compared to his Christian faith and all other previous cosmologies, because it was based on Dar-win's theory of evolution. As a scientist, he could claim the author-ity of the best factual knowledge of his day. As a priest, his writing had an inspirational quality that went beyond a purely scientific treatise. Alone, science only tells us what *is* and not what *ought to be*. To get from "is" to "ought," we must combine facts with values. It is a fact that torture causes pain, but we need the value that causing pain is wrong before we can conclude that we ought not to torture. Teilhard described the Omega Point as not just a theoretical possi-bility but a sort of heaven on earth worth pursuing with one's heart and soul.

Today, over half a century after his death, Teilhard remains widely read for his spiritual quality but has been largely forgotten by scientists. That's a pity, because *The Phenomenon of Man* was sci-entifically prophetic in many ways. Living processes indeed rival non-living physical processes in shaping the planet earth and its atmosphere. As we are increasingly coming to realize, our species does represent a new evolutionary process—cultural evolution—that far surpasses cultural traditions in other species. This capacity for cultural evolution enabled our ancestors to spread over the globe, inhabiting all climatic zones and dozens of ecological niches. Then small-scale societies—"tiny grains of thought"—coalesced into larger and larger societies over the past ten thousand years. Human activities now rival other living processes and non-living physical processes in shaping the earth and atmosphere, exactly as Teilhard said. It's even true that the Internet and other technological marvels are capable of furnishing the earth with a planetary brain.

This book can be seen as an updated version of *The Phenomenon of Man*. It is fully scientific, based on the best of our current knowledge about evolution, which has grown vastly more sophisticated since Teilhard's day. It also unabashedly goes beyond what *is* to provide a blueprint for what *ought to become*. Modern evolutionary theory shows that what Teilhard meant by the Omega Point is achievable in the foreseeable future. However, the same theory shows that its achievement is by no means certain. The reason is that evolution is *both* the solution *and* the problem. The harmony and order that we associate with the word "organism" indeed has a movable boundary that can be expanded to include biological ecosystems, human societies, and conceivably the entire earth. Special conditions are required, however, and when these conditions are not met, evolution takes us where we don't want to go. There is no master navigator for our journey. We must be the navigators, consciously evolving our collective future, and without the compass provided by evolutionary theory, we will surely be lost.

This View of Life

This View of Life

Whatever you think you currently know about evolution, please move it to one side to make room for what I am about to share in the pages of this book. I think you'll find that my argument doesn't fall into any current category. Politically it isn't left, right, or libertarian. It's not anti-religious and it enables us to think more deeply about religion than ever before. Above all, it moves us in the direction of sustainable living at all scales. Who doesn't want to improve their personal well-being; their families, neighborhoods, schools, and businesses; their governments and economies; and their stewardship of the natural world? These goals are within reach—but only if we see the world through the lens of the right theory.

To begin, we need to do a bit of clear thinking on what science is. It is commonly portrayed as a contest between theories that are based on a common stock of observations. First we see and then we theorize. Theories that do the best job of explaining the observations are accepted, only to be challenged by another round of theories, and so on, bringing our knowledge of the world closer to reality.

The problem with this view of science is that the common stock of observations is nearly infinite. We cannot possibly attend to everything, so a theory—broadly defined as a way of interpreting the world around us—is required to tell us what to pay attention to and what to ignore. We must theorize to see. A new theory doesn't

Albert Einstein

just posit a new interpretation of old observations. It opens doors to new observations to which the old theories were blind.

Albert Einstein understood this point when he wrote, "It is the theory that decides what we can observe." He was corresponding with his colleague Werner Heisenberg about electron orbits inside atoms. There was no way to directly observe electron orbits at the time, and Heisenberg thought it prudent to theorize on the basis of what can be observed. Einstein understood that theorizing about entities that cannot yet be seen can lead to useful predictions about what can be seen, but which had previously gone unnoticed.

Charles Darwin experienced the blindness that comes from lack of the right theory as a young man on a fossil-hunting expedition with his professor Adam Sedgwick. The valley in Wales that they visited had been scoured by glaciers and therefore had no fossils. The evidence for glaciers lay all around them—the scored rocks, the perched boulders, the lateral and terminal moraines, all typical of a glaciated landscape. Yet Darwin and Sedgwick were blind to the evidence because the theory that vast sheets of ice had once covered much of the northern hemisphere had not yet been proposed. They didn't know what they should have been looking for. Darwin commented in his autobiography that "a house burnt down by fire did not tell its story more plainly than did this valley. If it had still been filled by a glacier, the phenomena would have been less distinct than they are now."

Darwin went on to contribute his own eye-opening discoveries with his theory of natural selection. The theory is amazingly simple: 1) Individuals vary; 2) Their differences often have consequences for survival and reproduction; 3) Offspring resemble their parents. Given these three conditions, populations will change over time. Traits that contribute to survival and reproduction will become more common. Individuals will become well adapted to their environments.[1]

The theory of natural selection is so simple and rests upon such firm assumptions that it seems obvious in retrospect. As Thomas Huxley famously remarked upon encountering it for the first time, "How stupid of me not to have thought of that!" Nevertheless, for those who first started to explore the implications, it was as if the scales had fallen from their eyes. Wherever they looked—the fossil record, comparative anatomy, the geographical distribution of species, and the many wonderful contrivances that adapt organisms to their environments—they found confirming evidence. In the contest of theories, the biblical account of creation didn't stand a chance. By 1973, the geneticist Theodosius Dobzhansky could

Young Charles Darwin

declare that "nothing in biology makes sense except in the light of evolution."[2]

MY STORY

I was a graduate student at Michigan State University in 1973 and my personal experience can help to explain what Dobzhansky meant by his imperious-sounding proclamation. As someone who loved the outdoors and aspired to be a scientist, I decided to become an ecologist so I could study animals in their natural environments. In keeping with the old joke about experts knowing more and more about less and less until they know everything about nothing, my research was focused on the feeding behavior of a tiny aquatic crustacean called a copepod. Even for this esoteric subject, the possibilities were endless. Copepods might select their food in any number of ways and a theory was needed to narrow the possibilities. Evolutionary theory predicts that copepods should feed in ways that enhance their survival and reproduction. This could mean maximizing the amount of energy harvested, feeding in a way that avoids being eaten by predators, or other possibilities that depend upon the details of the environment. No theory leads directly to the right answer. The best that a theory can do is to narrow the field of possibilities. In this case, I predicted that copepods might selectively graze on larger algae rather than harvesting algae without respect to size, which would increase their rate of energy intake. My prediction turned out to be correct and resulted in my first publication in 1973.[3] I had added one small but solid brick to the edifice of scientific knowledge. I couldn't vouch for Dobzhansky's claim for all of biology, but I could testify that evolutionary theory had helped to make sense of my little corner.

Later that year I traveled to Costa Rica to attend a course in tropical ecology run by the Organization for Tropical Studies (OTS), a network of field stations run by a consortium of universities that is still going strong after fifty years. It was a life-changing experience. Anyone who loves nature is thrilled by the tropics, but we were seeing all of those wonderful plants and animals through

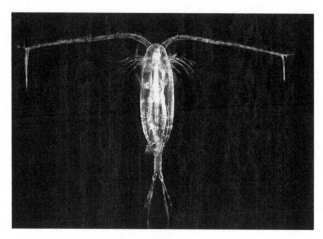

A copepod, my first scientific love

the lens of evolutionary theory—the same lens that informed my esoteric research. I realized that I didn't need to spend my life studying copepods. I could pick any creature or topic that interested me and quickly start asking intelligent questions based on the logic of evolutionary theory. It was the opposite of the old joke about experts knowing more and more about less and less. Becoming an expert in evolutionary theory was like receiving a passport to the study of all aspects of life.

Ever since, I have used evolutionary theory to study a multitude of creatures and topics. I have also witnessed my field of biology become ever more sophisticated in observational techniques. The modern biologist is like Darwin with superhuman powers of observation.[4] He or she can catalog entire genomes and track the patterns of gene expression (epigenetics); can trace neural pathways inside the brain; can monitor the movement of animals via satellite; can measure climate change in the distant past with a high degree of accuracy; can experiment with evolution in the laboratory using microbes that can be frozen and brought back to life to compare with their own descendants.

These technological marvels bring the common stock of observations well beyond anything imagined in Darwin's day. The role of evolutionary theory in making sense of all this information is more important than ever before. Dobzhansky's 1973 statement that nothing in biology makes sense except in the light of evolution has withstood the test of time. Yet for many people the word "biology" conjures up a different set of associations than words such as "human" or "culture." To proceed further, we need to expand the scope of what we consider biology.

HOW ABOUT US?

Darwin was convinced that his theory could explain the length and breadth of humanity, in addition to everything else more typically associated with biology. He observed and kept notes on his children with the same discerning eye that he observed barnacles and orchids. Following upon *On the Origin of Species*, he developed his thoughts at length in *The Descent of Man* and other works.

Yet, as evolutionary biology became a branch of science, the study of humanity did not proceed along the same track. The problem was not just the collision with religious belief that remains with us today. Some people who were fully comfortable with a naturalistic conception of the world still had an allergic reaction to evolutionary theory in relation to human affairs. As early as the 1870s, the threat that they perceived was given a name: social Darwinism.

According to most people's conception of social Darwinism, the haves and have-nots of society are equivalent to the fit and unfit of evolutionary theory. It is nature's way for the fit to replace the unfit. Interfering with the process would degrade the species and lead to the collapse of society. It is not selfish for the fit to replace the unfit; it is a moral imperative. Policies that flow from this logic include laissez-faire capitalism, withholding welfare from the poor, forced sterilization, and genocide.

After the excesses of the Gilded Age, the eugenics policies enacted in America and Britain, and the genocidal horrors of World War II, the idea of using evolutionary theory to formulate public policy became unthinkable. The stigma carried over to the aca-

MR. BERGH TO THE RESCUE.

THE DEFRAUDED GORILLA. "That *Man* wants to claim my Pedigree. He says he is one of my Descendants."
Mr. BERGH. "Now, Mr. DARWIN, how could you insult him so?"

The cartoonist Thomas Nast makes fun of Darwin's theory that we are descended from the apes.

demic disciplines classified as the social sciences and humanities. While these areas of study developed into sophisticated bodies of knowledge, they largely avoided engaging with evolutionary theory.[5] Most humanist scholars were happy to accept Darwin's theory for the study of the rest of life, our physical bodies, and a few basic instincts such as to eat and have sex, but insisted that our rich behavioral and cultural diversity operated according to a different set of rules.

AN INTELLECTUAL APARTHEID

The publication in 1975 of the book *Sociobiology: The New Synthesis* by Edward O. Wilson reveals the intellectual apartheid that was in place at the three-quarter mark of the twentieth century. The thesis of Wilson's book was that evolution provides a single theoretical framework for understanding the social behavior of all species, from microbes to humans. It was regarded as a triumph except for the final chapter on humans, which created an uproar—complete with accusations of fascism. Someone even dumped a pitcher of water onto Wilson's head at a conference because they were so offended by the perceived implications of his arguments.

Thus, at the same time that Dobzhansky could declare that nothing in biology makes sense except in the light of evolution, evolutionary theory was vigilantly being kept out of the social sciences and humanities. A full century had been lost. Terms such as evolutionary psychology, evolutionary anthropology, and evolutionary economics weren't coined until the 1980s, signaling a renewed attempt to rethink these disciplines from an evolutionary perspective—and even these had the whiff of scandal about them.

Today, I'm happy to report that the academic study of humans from an evolutionary perspective is getting back on track. Open any issue of a premier academic journal such as *Science, Nature,* and the *Proceedings of the National Academy of Sciences,* or any of hundreds of specialized academic journals, and you will find articles that view the length and breadth of humanity through the lens of evolutionary theory as if it is uncontroversial—just as biologists study the rest of life. However, there is a lot of catching up to do. Evolution is still taught primarily as a biology course at most colleges and universities. Many professors in the social sciences and humanities, who received no evolutionary training during their student days, are unlikely to modernize their views. Even most evolutionary biologists still associate evolution with genetic evolution, ignoring the fact that cultural change is also an evolutionary process—the key insight of Teilhard almost a century ago. The scientists and scholars who are back on track might number in the thousands, but this is a tiny fraction of the worldwide academic community.

When we move from the world of academia to the world of politics and public policy formulation, the situation gets even worse. Evolution is still a toxic word in the political arena. In addition to losing the creationist vote, it still invokes the specter of social Darwinism in the minds of many people. Even politicians and policy experts who are open-minded about evolution typically don't have the slightest idea how evolutionary theory might inform their particular corner of the policy world.

PILGRIM'S PROGRESS

Let's take stock as we begin our journey. The claim of this book is that a firm knowledge of evolutionary theory, which includes our own species, is required to solve the problems of our age. Yet, our current knowledge of humanity and our many attempts to improve our circumstances are to a large extent *pre-Darwinian*.

This claim will probably sound strange to most readers. I am not referring to the collision between evolutionary theory and religious beliefs. I am not even referring to a conscious denial of evolutionary theory for any reason. These topics can be set aside, at least for the moment. A person can fully endorse a naturalistic view of the world, including the fact of evolution, and still be pre-Darwinian.

To explain what I mean, let's return to Darwin and Sedgwick looking for fossils in the Welsh valley scoured by glaciers. A *particular* theory was required to make sense of what lay around them. No other theory would do. Our current knowledge of humanity and our many attempts to improve the human condition with public policy consist of hundreds of theories and rationales that are too informal to call theories but nevertheless draw our attention to some possibilities and blind us to others. Most of them are "local" theories, which means that they attempt to explain a limited range of phenomena without pretending to have more general explanatory scope. They are seldom related to each other or to any general theoretical framework, much less evolutionary theory. This is radically different than the conventional biological sciences, where all topics are approached from a single theoretical perspective.

Scientists, scholars, politicians, and policy experts might *think* that their theories and rationales are consistent with evolutionary theory, but in the absence of explicit inquiry there is no way to know. When these topics are approached from an evolutionary perspective, massive problems in the ways they have been conceptualized are often revealed. And new possibilities emerge that in retrospect are so obvious that a house burnt down by fire would not tell its story more plainly.

Dozens of examples will be provided in the pages of this book, but consider the economics profession as a foretaste. It includes many schools of thought, but the dominant school is inspired not by Darwin's theory of evolution but by nineteenth-century physics, as if an economic system can be predicted with the same kind of mathematical precision as the orbits of planets around the sun in our solar system.[6] Even though the Norwegian-American economist Thorstein Veblen wrote an article titled "Why Is Economics Not an Evolutionary Science?" in 1898,[7] the term "evolutionary economics" was not coined until the 1980s,[8] reflecting the more general intellectual apartheid that I have already recounted. Today, evolutionary economics is a tiny heterodox school of thought with almost no influence on economic policy. One prominent economist who gets it right is Robert Frank, who predicts that within a hundred years, Darwin and not Adam Smith will be regarded as the father of economics.[9] If he is correct, then it will have taken over two centuries to complete the Darwinian revolution for this topic area, and we are still only a little more than halfway there!

Economics is one of a very few disciplines in the social sciences and humanities that even pretends to have a unifying theoretical framework. Most other disciplines, such as political science, sociology, history, psychology, and education, are nothing more than a constellation of schools of thought that are poorly related to each other, much less to evolutionary theory.

TOWARD AN EVOLUTIONARY WORLDVIEW

Completing the Darwinian revolution therefore requires a massive reset in our understanding of humanity, which must take place at a timescale of years, not decades. We need not just a theory that states what *is*, but a worldview that informs how we *ought* to act, while remaining fully within the bounds of scientific knowledge. Here is a preview of the voyage that we will be taking in the pages of this book.

We must begin by confronting the dark past of social Darwinism. Is it true that Darwin's theory unleashed a plague of toxic policies justifying social inequality and is inherently more dangerous than other theories? The answer to this question is more complex and intriguing than you might think.

Next, I show how evolutionary theory provides a toolkit for making sense of any living process—a fact that is already accepted for biology. I use the word "toolkit" to invoke the mindset of a plumber or carpenter: someone who arrives at a site, sizes up the job, and pulls out the right tools to get the job done. Anyone can become adept at this kind of work—including you—and I can get you started within a single chapter.

Once we are equipped with a toolkit, we can get right to work addressing problems that everyone wants to solve, such as our physical and mental health and optimal ways to raise our children. Along the way, we will see that there is no dividing line between "biology," "human," and "culture." The same conceptual tools are useful for making sense of everything associated with these three words.

The next step in our journey turns what religious believers call "the problem of evil" on its head. Their problem is to explain how everything associated with the word "evil" can exist in a world created by an all-powerful and beneficent god. The problem of the evolutionist is to explain how everything associated with the word "goodness" can evolve in a Darwinian world. Modern evolutionary theory tells us that goodness *can* evolve, but only when special conditions are met. That's why we must become wise managers of evolutionary processes. Otherwise, evolution takes us where we don't want to go. Along the way, we'll see how the same toolkit can make sense of examples as diverse as cancer, psychopathic chickens, and the very nature of human morality.

If you're like most people, you probably regard evolution as such a slow process that it stands still for the timescales that matter most to us. That's not necessarily true even for genetic evolution, which can take place in a single generation. Once we appreciate that evolution goes beyond genetic evolution, however, we can begin to use our toolkit to understand the fast-paced cultural changes swirling all around us and even within us as actively evolving entities in our own right.

The problems that require our attention exist at all scales, from individual well-being to a sustainable planet. Evolutionary theory distinctively identifies the *small group* as a fundamental unit of human social organization, required for both individual well-being and effective action on a larger scale. Nearly every reader of this book can become happier as an individual and more engaged as a citizen by creating and joining small groups. Yet, small groups don't automatically work well, and even when they do, they can become part of the problem higher up the scale, such as a street gang, a terrorist cell, a predatory corporation, or a rogue nation. We must consciously seek to create small groups that benefit individuals as well as society as a whole.

It is one thing for a species to be *well adapted* to its environment and another for it to be *adaptable* to environmental change. The same goes for human cultures, and almost no existing culture is adaptable enough to keep pace with our ever-changing world. Conscious evolution requires the construction of a new system of cultural inheritance capable of operating at an unprecedented spatial and temporal scale. This will be a formidable task, but evolutionary theory does provide the tools to get the job done.

Our journey ends with a reflection on how the secular imagination and the religious/spiritual imagination can converge on the conscious evolution of our collective future. It is striking how these two imaginations often seem at odds with each other, yet both arrive at the same conclusion: that the concept of "organism" has a movable boundary, which must be expanded to solve the problems of our age. It's as if two separate languages are being spoken, each of which apprehends the same reality in its own way but which are mutually unintelligible to most of their speakers. In my own journey, I have learned to speak both languages and to appreciate how they both add value to consciously evolving our collective future. I hope that by the end of this book, you will become bilingual as well.

Dispelling the Myth of Social Darwinism

Before an evolutionary worldview can be established in a positive sense, we must look at the troubled history of social Darwinism. If grave social injustices were committed in Darwin's name, such as withholding welfare from the poor, forced sterilizations, racial discrimination, and outright genocide, then this is truly dangerous territory. But this rendering of social Darwinism is largely a myth and the true history is far more interesting and complex. Darwin's theory, properly understood, is centered on cooperation, and Darwin and others were clear about this from the beginning.

NOW AND THEN

Let's begin with the modern meaning of the phrase "social Darwinism." Type it into your favorite search engine and you'll see thousands of references to the strong gaining at the expense of the weak, as in this cartoon. Former U.S. president Barack Obama was fond of using the phrase in attacks on his Republican opponents, as in a 2012 speech when he called a budget proposed by the U.S. representative Paul Ryan "thinly veiled social Darwinism."[1]

Two facts can quickly be established about these usages. First, the phrase "social Darwinism" (or "social Darwinist") is almost

Today's meaning of social Darwinism

always used as a pejorative to describe someone else's position. People don't call themselves social Darwinists. They are called social Darwinists by their opponents. Second, the people who are labeled social Darwinists seldom actually use Darwin's theory of evolution to justify their position. The idea that Paul Ryan and his associates would use Darwin's theory to argue in favor of their budget is downright funny, since a vocal portion of this constituency denies evolution altogether!

So many books and articles stretching back to the nineteenth century have been digitized that it is possible to comprehensively search for the phrase "social Darwinism" to trace its origins and history of use. Geoffrey M. Hodgson, a contemporary scholar of Darwinism in relation to human affairs, has performed the Herculean task of tracking down every usage of the phrase in JSTOR, an electronic database of academic journals.[2] He shows that the phrase began to be used during the late nineteenth century and the context was exactly the same as in the present. As he puts it: "The label was used primarily by leftists to pin upon their opponents."

Hodgson also discovered that the phrase was hardly used at all until popularized by *Social Darwinism in America*, a book published by the historian Richard Hofstadter in 1944. This means that during the period when many of the abuses associated with social Darwinism occurred, such as eugenics policies in the United States and Britain and Nazi policies leading to the Holocaust in World War II, the term "social Darwinism" was rarely used by either proponents or opponents of the policies.

To summarize, from the very beginning and throughout its history, the term "social Darwinism" has been used as a pejorative to describe policies that justify the strong taking from the weak in various forms. People hardly ever call themselves social Darwinists and those who are accused almost never actually use Darwin's theory to justify their position. To a remarkable extent, Darwin's theory stands falsely accused.

Since the term "social Darwinism" proves to be such an unreliable guide, let's look back on some of the major protagonists of early evolutionary theory to see what they thought about the role of competition in the betterment of society, including some of Darwin's predecessors, Darwin himself, and those who were influenced by him.

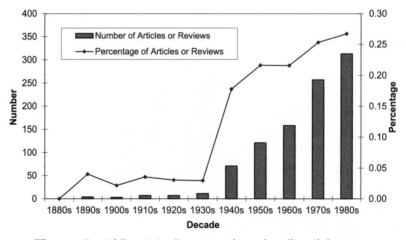

The term "social Darwinism" was scarcely used at all until the 1940s.

THE CLERIC

Thomas Robert Malthus (1766–1834) was a British cleric and scholar, whose essay on population growth helped both Darwin and Alfred Russel Wallace to independently formulate their theories of natural selection.[3] Malthus observed that populations will inevitably grow to exceed the capacity of resources available to sustain them and therefore will be checked by famine and disease. As a cleric, he concluded that this situation must be divinely imposed to teach virtuous behavior. Misery was part of God's plan and should not be impeded by well-meaning but shortsighted attempts to improve the lot of the poor. God's plan should be allowed to take its course. Needless to say, this was a departure from traditional Christian doctrine, which mandates extending charity to the poor.

Malthus was not totally heartless in the development of his thesis. For example, he thought that moral constraint on the part of the poor was superior to famine and disease, but in either case he argued against the Poor Laws, which was the government's social

Malthus thought that misery was part of God's plan to teach virtuous behavior.

welfare system at the time. Malthus's work, in addition to inspiring the theory of natural selection, influenced the early development of economic theory.

THE POLYMATH

Herbert Spencer (1820–1903) was regarded as an intellectual giant during Darwin's day and had his own theory of evolution that preceded Darwin's.[4] Today his star has almost completely faded and none of his ideas about evolution are taken seriously. To understand such a spectacular rise and fall, we must appreciate what Spencer meant to his peers. He represented an effort to create an all-embracing cosmology that could explain everything on the basis of natural laws and could be discovered by scientific investigation. In short, Spencer was offering a substitute for conventional religion that was centered on science. The French philosopher and polymath Auguste Comte (1798–1857) had the same expansive goal to create a "Religion of Man," but Spencer attempted to organize his cosmology around a single law of progressive development

Spencer championed women's rights and opposed imperialism, but he also opposed welfare for the poor.

that he called the principle of evolution, which could be applied to all branches of knowledge. It is easy to imagine the appeal of this project for Enlightenment thinkers, however doomed its prospects were over the long term.

Spencer was a political radical as a young man, although he became more conservative as he aged. He was against the power of the aristocracy and in favor of a voting privilege for women and even children. He opposed British imperialism and military conquest throughout his life. Nevertheless, his "progressive" views led him to oppose extending social welfare to the poor, along with most other state-controlled activities. For Spencer, humanity was evolving toward perfection without the help of governments, which only got in the way. There is a clear family resemblance between Spencer's views and the laissez-faire economic policies of today. Spencer departed from Malthus in centering his cosmology on science rather than religion, but he shared Malthus's view that "survival of the fittest" (a phrase that Spencer coined) played an essential role in the perfection of humanity and therefore should not be impeded by humanitarian impulses and especially not by state-controlled programs.

THE FOUNTAINHEAD

Now we are in a position to consider Darwin's views on the role of "survival of the fittest" in human society.[5] He was progressive by the standards of his day but still very much a product of Victorian culture. He was vehemently against slavery and regarded all races as part of a single human species, but he also regarded European culture as superior and could not conceal his distaste for some of the people that he encountered on his voyage on the *Beagle*. He took it for granted that men were intellectually superior to women. On eugenics, he assumed that selective breeding practices would work for humans as well as for domesticated plants and animals, but nevertheless counseled against human eugenics because dictating who could breed would violate feelings of sympathy, which he regarded

Darwin was
passionately against
slavery but still
regarded European
culture as superior,
along with almost all
other Europeans at
the time.

as the glue that holds human society together. As he put it: "Selfish and contentious people will not cohere and without coherence
nothing can be effected."[6] Clearly, Darwin placed the emphasis on
cooperation rather than competition.

When the term "social Darwinism" began to be used, it did not
refer to Darwin's own views but to the laissez-faire views associated
with Spencer and Malthus that were already in place before Darwin made the scene. The epithet bore Darwin's name because his
theory of evolution was the most widely accepted and Spencer's had
fallen out of fashion. The few people who used the phrase weren't
as concerned about making scholarly distinctions between Darwin
and Spencer as they were about stigmatizing an opponent.

It was inevitable that others would speculate about the implications of Darwin's theory for public policy, even if Darwin remained
circumspect. Three notables were Francis Galton, Thomas Huxley,
and Peter Kropotkin.

THE COUSIN

Francis Galton (1822–1911) was Darwin's half cousin and a prolific scientist in his own right.[7] He contributed to the early study of statistics in addition to other achievements. He was politically progressive for his day, and his scientific orientation toward all topics led him to study the efficacy of prayer by comparing the longevity of clerics with men of other professions (there was no difference).[8] He is the person who coined the word "eugenics" to promote selective breeding for the improvement of humans, using the same principles that had been employed for generations to improve domesticated plants and animals.

By his own account, Galton was inspired by Darwin's *Origin of Species* to investigate the heritability of human traits, including everything from fingerprints and facial morphology to mental attributes. His book *Hereditary Genius*, published in 1869, claimed to provide evidence for the inheritance of human abilities. It made sense to Galton for government to implement programs that favored

Galton favored a strict meritocracy, arguing against hereditary privilege and for accepting talented immigrants.

the reproduction of people with abilities. The particular plan that he advocated was by today's standards a bizarre mosaic of liberal and conservative ideas. Everyone should receive a first-class education and entrance to professional life. Income should be derived from work rather than inheritance so that everyone has a chance to demonstrate their innate abilities. After the state has provided a level playing field, it should provide unisex monasteries for the weak to live out their lives in comfort without reproducing. Immigrants should be screened for their abilities and the best should be naturalized.

THE BULLDOG

Thomas Henry Huxley (1825–1895) was born into a middle-class family that fell upon hard times.[9] He had only two years of formal schooling but nevertheless educated himself to become one of the foremost biologists of his day, specializing in comparative anatomy. Like Darwin, he embarked upon a sea voyage as a young man (as a surgeon's mate on the HMS *Rattlesnake*) and made biological inves-

Huxley ridiculed eugenics as a "pigeon fancier's polity."

tigations that earned him a reputation back home. He was elected a fellow of the Royal Society at the age of twenty-six. He was skeptical about the evolutionary theories that preceded Darwin but became convinced about the fact of evolution by *On the Origin of Species*, even though he remained less certain about the centrality of natural selection (in general, the concept of identity by descent was accepted more quickly by Darwin's peers than natural selection as the driving force of evolution).[10] Combative, eloquent, and extroverted, Huxley described himself as "Darwin's bulldog" in defending Darwin against his critics.

Huxley's views on evolution in relation to human affairs were expressed in "Evolution and Ethics," an essay written a few years before his death and still in print today. Huxley rejected Galton's eugenic vision, lampooning it as a "pigeon fancier's polity," because no one is smart enough to identify the abilities of people, especially at a young age. Huxley must have had in mind his own experience as someone who was not born into privileged society when he wrote: "I doubt whether even the keenest judge of character, if he had before him a hundred boys and girls under fourteen, could pick out, with the least chance of success, those who should be kept, as certain to be serviceable members of the polity, and those who should be chloroformed, as equally sure to be stupid, idle, or vicious."[11]

Nevertheless, Huxley thought that some kind of selection was required to create a moral society. The natural world is not moral and human society left unmanaged isn't either. Our evolved passion for pursuing our own self-interest would lead to the destruction of society unless somehow constrained. Fortunately, in addition to our self-aggrandizing instincts, we also have instincts that can be used to create a moral society, not least our passion to maintain a good reputation: "I doubt if the philosopher lives, or ever has lived, who could know himself to be heartily despised by a street boy without some irritation."[12] We therefore need to work with what evolution has provided us to cultivate a moral society that works for the benefit of all, in the same way that we cultivate a garden.

Huxley was among the most gifted writers of his age, and his critique of the "survival of the fittest" worldviews of Malthus, Spencer, and Galton (bearing in mind that each of these men formulated different versions, as we have seen) is worth savoring.

> It strikes me that men who are accustomed to contemplate the active or passive extirpation of the weak, the unfortunate, and the superfluous; who justify that conduct on the ground that it has the sanction of the cosmic process, and is the only way of ensuring the progress of the race; who, if they are consistent, must rank medicine among the black arts and count the physician as mischievous preserver of the unfit; on whose matrimonial undertakings the principles of the stud have the chief influence; whose whole lives, therefore, are an education in the noble art of suppressing natural affection and sympathy, are not likely to have any large stock of these commodities left. But, without them, there is no conscience, nor any restraint on the conduct of men, except the calculation of self-interest. . . .[13]

Huxley's views were similar to Darwin's on this point, but far more forcefully expressed. The bulldog indeed had a fierce bite.

THE ANARCHIST

Peter Kropotkin (1842–1921) was a Russian prince who became a prominent anarchist, arguing for a communist society free from central government.[14] Kropotkin's father owned large tracts of land inhabited by over a thousand serfs. The son's proclivities toward equality emerged early, and at the age of twelve he stopped using his princely title. Like Darwin and Huxley, Kropotkin traveled widely (although by land instead of sea) and conducted scientific investigations, centered on geology but including observations of animals and the many indigenous cultures making up the Russian Empire. He was imprisoned repeatedly for his revolutionary activities and also spent periods in exile in Britain, Switzerland, and France. He returned to Russia as a hero in 1917 but became disillusioned when the Bolsheviks seized power.

Kropotkin regarded mutual
aid as the main survival tool of
humans and other animals.

Kropotkin's main contribution to evolutionary theory was his book *Mutual Aid: A Factor in Evolution,* which was published in 1902. He argued that the emphasis on competition among individuals was misplaced and an artifact of British capitalism. In reality, most species live in groups whose members provide mutual aid to each other in their common struggle against the environment. Kropotkin also claimed that indigenous human societies were primarily cooperative and that this form of cooperation could provide a model for modern society without the need for a strong central government. It is interesting how the views of Kropotkin converge with those of Spencer on this libertarian point, despite profound differences in other respects.

A RAIN FOREST OF OPINIONS

Six major theorists—Malthus, Spencer, Darwin, Galton, Huxley, and Kropotkin—each with their own views on the role of competition in the betterment of human society. Malthus justified his views in terms of religion and Spencer in terms of a secular cosmology that used the word "evolution" but otherwise bore little resemblance to Darwin's theory. Spencer's views on social welfare fit the stereotype of social Darwinism, but not his views on aristocracy, imperialism, and military conquest. Galton's views on eugenics fit the stereotype of social Darwinism but not his views on providing a level playing field. Malthus, Spencer, and Kropotkin argued against "big government" while Galton expected his eugenic utopia to be orchestrated by the state. Malthus, Spencer, and Galton saw competition as a source of societal improvement, but Darwin, Huxley, and Kropotkin emphasized the corrosive effects of unbridled self-interest and the essential role of cooperation.

What are we to make of all this variety? Evolutionary theory meant different things to different people, which is only to be expected when an important new idea starts to interact with one's current worldview. As for these six major figures, so also for thousands of other people as Darwin's theory started to become known and discussed around the world.[15] There was a veritable rain forest of opinions in the discussions surrounding early evolutionary theory. Decades would be required to sort among them, just as decades were required to sort out the implications of Darwin's theory for the biological world. But with respect to human society, that process was largely halted by the false view that Darwin's theory leads to nothing but pernicious outcomes.

HITLER AND DARWIN

One of the most common claims about Darwinism is that it contributed to the genocidal horrors of Nazi Germany. Before examining the evidence, what should we conclude if this claim turns out to be true? Would it mean that we should avoid evolutionary theory in the formulation of public policy?

Even a brief reflection should convince you that the answer to this question is "no." Almost anything that can be used as a tool can also be used as a weapon. Do we avoid physics because it can lead to the atomic bomb? Do we avoid chemistry because the Nazis used chemicals in their gas chambers? Do we blame Mendel for misuses of genetics? The entire concept of making a scientific theory and its originator morally culpable for misuses of the theory is deeply misguided.

That said, there is no evidence that Hitler or his associates were influenced by Darwin, either directly or indirectly through Darwin's German colleague and friend, Ernst Haeckel. The distinguished historian of science Robert Richards has clearly demonstrated this point in his essay "Was Hitler a Darwinian?"[16]

Focusing on Darwin ignores the deep roots of racism in general and anti-Semitism in particular in European thought. It's hard to find a major figure who didn't arrange humans into a hierarchy of races with Europeans at the top. Carolus Linnaeus (1707–1778), the great classifier of plants and animals, divided *Homo sapiens* into four categories, each with their own temperaments: American (copper-colored, choleric, regulated by custom), Asiatic (sooty, melancholic, and governed by opinions), African (black, phlegmatic, governed by caprice), and European (fair, sanguine, and governed by laws). Since Linnaeus was a creationist, he assumed that these differences were divinely ordained.

Darwin, himself, took a hierarchy of races for granted but wrote almost nothing about Jews. Haeckel did include Jews in his racial hierarchy but ranked them at the same level as Germans and other Europeans. In an interview conducted in the 1890s, Haeckel stated: "I hold these refined and noble Jews to be important elements in German culture. One should not forget that they have always stood bravely for enlightenment and freedom."[17] So much for Haeckel's influence on Hitler!

Hitler and his associates drew upon books such as the four-volume *Essay on the Inequality of the Human Races*, by Joseph Arthur, comte de Gobineau (1816–1882), and *The Foundations of the Nineteenth Century*, by Houston Stewart Chamberlain (1855–1927), who was part of a cult that formed around the composer Richard Wagner (1813–1883). Gobineau developed his thesis that race mixing was responsible for the decline of human societies prior to Darwin and disdainfully dismissed Darwin's theory. Chamberlain dismissed as "pseudo-scientific fantasy" Darwin and Haeckel's argument that humans were descended from apelike ancestors.

The connection between Hitler and Chamberlain is plain to see. Hitler read Chamberlain's book sometime between 1919 and 1921. They met when Hitler visited Wagner's home and shrine in 1923. Chamberlain was among the first of Hitler's supporters, and Hitler attended Chamberlain's funeral in 1927. In contrast, the historical record includes no mention of Hitler discussing Darwin's theory of evolution until 1942, when he said: "Whence have we the right to believe that man was not from the very beginning what he is today? A glance at nature informs us that in the realm of plants and animals alterations and further formation occur, but nothing indicates that development within a species has occurred of a considerable leap of the sort that man would have to have made to transform himself from an apelike condition to his present state."

As Richards remarks after quoting this passage, could any statement rejecting Darwin's theory of evolution for the human case be more explicit? Elsewhere in his essay, Richards compares the search for connecting Darwin to Hitler as similar to seeing animal shapes in clouds or the game of six degrees of separation, which could equally be played to connect Hitler to Aristotle or Jesus. When it comes to influencing the genocidal horrors of Nazi Germany, Darwin's theory is clearly not responsible.

DEWEY AND DARWIN

If a contest were to be held for the opposite of the stereotypic social Darwinist, the winner might well be John Dewey (1859–1952), the American philosopher, psychologist, educator, and social reformer.[18] Yet, Darwin's influence on Dewey is as plain to see as Chamberlain's influence on Hitler. The reason that no one calls Dewey a social Darwinist is because the term is used entirely as a pejorative, giving the misleading impression that anyone who falls under the spell of Darwin is going to end up thinking like Hitler.

Dewey represented the philosophical tradition of Pragmatism, which originated during the late nineteenth and early twentieth centuries among a small group of American intellectuals, including Oliver Wendell Holmes, Charles Sanders Peirce, and William

James. Their stories are beautifully told by Louis Menand in his Pulitzer Prize–winning book *The Metaphysical Club*. They realized that if the human mind is a product of natural selection, then knowledge must be based on its practical consequences rather than a God-given apprehension of objective reality. This was a radical new approach to epistemology, the branch of philosophy devoted to understanding the nature of knowledge.

By the time John Dewey attended college at the University of Vermont and graduate school at Johns Hopkins University, Darwin, Spencer, and the Pragmatists were part of his education. Although Spencer is remembered today for his laissez-faire attitudes toward the poor, his broader philosophical view emphasized the intimate relationship between organisms and their environments. In fact, the word "environment" was seldom used prior to the mid-nineteenth century, as strange as this might seem to us today. It was Spencer, not Darwin, who popularized its use. Darwin used terms such as "conditions" and "circumstances" and didn't start using the term "environment" until late in his career.

Dewey found it easy to embrace Spencer's emphasis on an

Dewey represents the philosophical tradition of Pragmatism, which was heavily influenced by Darwin.

organism-environment relationship while rejecting his laissez-faire attitude toward the poor. For Dewey, pragmatism meant "the applied and experimental habit of mind." In what he called an "evolutionary method," any given moral norm or mode of reasoning is "treated as an instrument of adjustment or adaptation to a particular environing situation." In other words, Dewey thought about evolution as a process taking place among beliefs and practices in the present, not just selection among genes that took place in the ancient past. This placed him close to Teilhard de Chardin and the modern concept of "niche construction" that will be discussed in this book.

Dewey's experimental approach led him to found the Laboratory School at the University of Chicago, which is still going strong. The idea that a school might serve as a laboratory for evolving best educational practices was a radical innovation at the time. He also worked closely with other second-generation Pragmatists such as Jane Addams, George Herbert Mead, and W. E. B. Du Bois, who are beloved as social reformers and champions of democracy and therefore never associated with the term social Darwinism.

Here is what Dewey had to say about Darwin in the title essay of his 1910 book, *The Influence of Darwin on Philosophy*:

> In laying hands upon the sacred ark of absolute permanency, in treating the forms that had been regarded as types of fixity and perfection as originating and passing away, the "Origin of Species" introduced a mode of thinking that in the end was bound to transform the logic of knowledge, and hence the treatment of morals, politics, and religion.[19]

For Dewey, "evolution" meant "change," with human agency having an essential role in the construction of a moral society.

THE CREATION OF A BOGEYMAN

As we have seen from Geoffrey Hodgson's meticulous analysis, the phrase "social Darwinism" was seldom used before it was popularized by the historian Richard Hofstadter (1916–1970) in his 1944 book *Social Darwinism in American Thought*. World War II was not a time for dispassionate scholarly analysis. As Hodgson puts it, "The skills of a great historian were deployed in the ideological war effort against fascism and genocide." Any view that promoted racism, nationalism, or competitive strife qualified as social Darwinism in Hofstadter's book.

Other forces were at work that created a separation between Darwin's theory in the biological sciences and the same theory in relation to human affairs. Darwin knew nothing about genes and formulated his theory in terms of variation, selection, and *heredity*—a resemblance between parents and offspring due to any mechanism. Once the science of genetics was born in the early twentieth century, however, genes came to be treated as the only mechanism of inheritance, as if there were no other way for offspring to resemble their parents. This is patently false, but the study of cultural inheritance mechanisms did not re-emerge within the biological sciences until the 1970s.[20]

Sociologists, for their part, had their own reasons for declaring independence from biology, and those reasons had nothing to do with the supposed evils of social Darwinism. In order to establish the study of society as a separate field of inquiry, they had to assert that it couldn't be "reduced" to biological or even psychological facts. The American sociologist Talcott Parsons used the term "social Darwinism" in this way, to refer to strictly biological influences, not including the kind of social constructivism that Dewey was happy to call evolutionary.[21]

In this fashion, evolution became widely regarded as a powerful theory for explaining the rest of life, the physical human body, and a few basic survival instincts—while having nothing to say about our rich behavioral and cultural diversity. Along the way, evolution and "biology" became associated with an *incapacity* for change (being stuck with our genes), with our capacity for change somehow residing outside the orbit of evolution.

The term "social Darwinism" helps to buttress this bizarre configuration of ideas in ways that are almost childish, once they are seen clearly. Here is how Geoffrey Hodgson puts it:

The woods can be dangerous. So we might tell children stories of woodland beasts or bogeymen, to warn them away from the forest. Similarly, prevailing accounts of "Social Darwinism" have been invented as bogeyman stories, to warn all social scientists away from the darkened woodland of biology. We are told that any use of ideas or analogies from biology in the social sciences is unsafe. We are warned not to stray into that biological zone, for terrible things will happen, as they surely happened before. But scientists should not be treated as children . . . Let us stop telling false histories, and henceforth call things by their proper names.

THE BIGGEST VICTIMS

Darwin's theory can and has been misused by some people. However, the same can be said for any theory and indeed for almost any tool of any sort. There is no evidence that Darwin's theory is especially prone to misuse, so using the term "social Darwinism" as a bogeyman has no place in serious scholarship or science. It's clear that policies classically considered social Darwinism actually have almost nothing to do with a proper reading and understanding of Darwin's theory.

Who are the biggest victims of this stigmatization? All of us. Evolutionary theory has made enormous progress within the biological sciences while remaining almost entirely excluded from the many branches of the human social sciences and humanities and their practical applications. Now that we have dispelled the myth of social Darwinism as a bogeyman, we can begin to explore how an evolutionary worldview can function in a positive sense to consciously evolve our collective future.

2

Darwin's Toolkit

One of the most disturbing realizations of the 2016 American presidential election—for Democrats and Republicans alike—was the fragility of truth. Most of us are exhorted to tell the truth throughout our lives. We know that we sometimes fail, since lying can be personally advantageous in a thousand different ways, but we also know that if everyone lied then something precious that we all rely upon—trustworthy information—would be lost. That's why most of us commit to telling the truth, expect to be punished when we lie, and punish other liars with zeal.

All that seemed to change with the 2016 American presidential election. Some people and organizations seemed to abandon the norm of truth-telling for themselves, wantonly generating "fake news" in order to achieve their own partisan ends. Worse, the societal response to lying seemed to break down. When fake news and other deceptions were brought to light—*nobody was punished*. In no time, fake news spread through the body politic like a cancer.

If you want a vision of hell on earth, you can't do much better than a world where no facts can be trusted. Against this background, the norms of scholarship and science begin to look downright heavenly. Scholars such as Geoffrey Hodgson and Robert Richards, whom we met in the last chapter, and the many scientists we will meet in this book are not saints—but they have a strong personal commitment to finding the truth. Even more important, they are held accountable by truth-telling norms that are strongly enforced. To communicate as a scholar or a scientist, you must document everything that you say and your work must be reviewed by your peers before it can be published. Even minor transgressions, such as failing to cite previous relevant work, can damage your reputation. Flagrant lying, such as falsifying data, results in expulsion. In this sense, scholars and scientists regard Truth as their religion.[1]

That said, we don't contemplate this often. We in academia and science are more like plumbers and carpenters on a day-to-day basis, showing up with a toolkit. We size up the job, reach for the appropriate tools, and get to work figuring out the facts of the matter. Some of the tools are physical objects, but there are also conceptual tools—certain ways of parsing information that are exceptionally insightful. In this chapter I will introduce you to the major conceptual tools in Darwin's toolkit, which were developed mostly for the study of genetic evolution in nonhuman species but which can also be used to study human genetic and cultural evolution, including the formulation of public policy.

NIKO TINBERGEN AND HIS FOUR QUESTIONS

Niko Tinbergen (1907–1988) was a Dutch biologist who shared the Nobel Prize in medicine with Konrad Lorenz and Karl von Frisch in 1973 for pioneering the field of ethology, or the study of animal behavior.[2] Back then, the idea that a behavioral trait such as aggression can evolve in the same way as an anatomical trait such as a deer's antlers or a physiological trait such as glucose metabolism was not widely accepted. In his effort to establish ethology as a branch of biology, Tinbergen called attention to four questions that must be addressed to fully understand any product of evolution. These questions are the most important tools in an evolutionary scientist's repertoire and are essential for understanding the arguments in this book.

Tinbergen pioneered the
study of animal behavior.

First, what is the *function* of a given trait (if any)? Why does it exist compared to many other traits that could exist? Second, what is the *history* of the trait as it evolved over multiple generations? Third, what is its physical *mechanism*? All traits, even behavioral traits, have a physical basis that must be understood in addition to their functions. Fourth, how does the trait *develop* during the lifetime of the organism? Recognizing these as separate questions and studying them in conjunction with each other form the foundational concepts in Darwin's toolkit.

For example, your hand is an adaptation that evolved by natural selection to grasp objects (its function). It evolved as part of the vertebrate lineage and is anatomically similar to the fins of fish (its history). Physically, it is composed of muscles, bones, tendons, and nerves put together in just the right way to grasp objects (its mechanism). It begins to appear as early as the fifth week of gestation (its development). A complete explanation of your hand requires answers to all four questions.

Fair enough, you might be thinking, but how can Tinbergen's four questions be used as tools in a policymaker's toolkit? Let's take a look at each question in more detail.

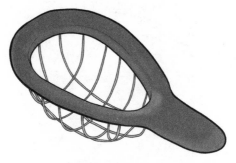

What is this for?

WHAT IS THIS FOR?

This gadget is clearly designed to do something, but what? Take a moment to think about it. Maybe it is a sieve, although most sieves don't have such a large mesh size. Maybe it's for sifting the poop from kitty litter, although I personally would like a longer handle for that job. Maybe it's a well-ventilated cup for protecting the private parts of male athletes. Wrong, wrong, wrong. It's an *avocado cuber*!

Whenever I play this game with an audience, a game show atmosphere develops. People who are normally shy about speaking in public eagerly raise their hands. There is a burst of laughter and satisfaction when they learn the right answer, which explains the details of the object so much better than the wrong answers. The size of the mesh and length of the handle are no longer a mystery. The overall shape (to conform to an avocado) and width of the rim (to push into the avocado) become important parts of the design, whereas before they might have appeared arbitrary.

Here's another puzzle for you. What is a snowflake designed to do? The one shown here is much more intricate than an avocado cuber. Nevertheless, the correct answer to my question is that a snowflake isn't designed to do anything. It wasn't made by people and it isn't a life form. Instead, it is the result of a physical process that takes place when water molecules crystallize around tiny particles in cold temperatures.

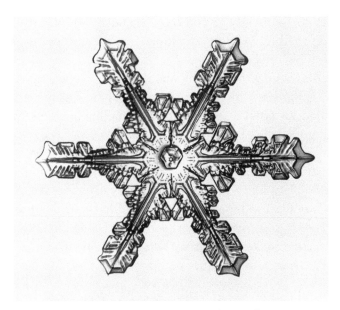

A snowflake has beautiful structure but no function.

To say that a snowflake has no purpose doesn't mean that it lacks value. Value is in the eye of the beholder, and I truly value the beauty of a snowflake, along with sunsets, cloudscapes, and the night sky. But none of these is designed for a purpose like a human implement or a life form that is a product of evolution. If you try to explain a snowflake or a physical process such as the weather as if they were designed for a purpose, confusion will result. If you try to explain a human-designed implement such as an avocado cuber or a biological adaptation such as a bird's wing as if it is not designed for a purpose, confusion will also result.

Here is another way to emphasize the crucial difference between an explanation that assumes functional design and one that doesn't. Suppose that you're walking up a mountain and you hear a crashing noise ahead of you. It's a falling boulder that will crush you if you don't move! You quickly calculate its trajectory and take note of objects that might change its path before you dart out of its way. That was a close call! Now suppose that the crashing noise is a charging grizzly bear rather than a boulder. If you thought about the bear the same way as you did a boulder, you would be dead. Avoiding the bear requires a completely different line of thought than avoiding the boulder.

Let's call the way we think about avocado cubers, bird wings, and charging bears "functional reasoning mode" and the way we think about snowflakes, the weather, and falling boulders "physical reasoning mode." All of us have a lifetime of experience reasoning in both modes, but we are also prone to misclassifications. We mistake a purely physical object such as a snowflake for a functionally designed object such as a bird's wing, or vice versa. We correctly classify an object as functionally designed but attribute the wrong function to it, like mistaking an avocado cuber for an athletic cup. When these mistakes happen, we become as blind to reality as Darwin and Sedgwick were bumbling around that Welsh valley looking for fossils.

THE PRE-DARWINIAN WORLDVIEW

Mistakes took place on a grand scale prior to Darwin's theory of evolution. According to the Christian worldview, the entire universe was created by a benign and all-powerful god. Everything from the tiniest insect to the stars in heaven plays a role in his grand plan. Even human suffering and the appearance of evil have a purpose that we must attempt to decipher. That's why Malthus interpreted famine and disease as divinely imposed to teach virtuous behavior. Based on what we know today, most of these errors were the result of confusing purely physical processes for designed processes or attributing the wrong functions to designed objects.

The emergence of science during the Enlightenment started to correct some of these errors but left others alone. Isaac Newton stunned the world with his ability to predict the motions of the planets with mathematical equations, but he also saw purpose and design in the heavens. For him and most other Enlightenment thinkers, scientific inquiry provided a new way to understand God by studying his creation. Scientific knowledge often disagreed with biblical accounts, which led to a conflict between Enlightenment thinkers and the established church. Even the scientists, however, assumed that their inquiry would reveal an orderly and harmonious universe, a giant clock where every part contributes to the working of the whole.

The concept of laissez-faire in economic theory and practice is a direct descendant of Christian and Enlightenment thinking prior to Darwin. Laissez-faire refers to a policy or attitude of letting things take their own course without interfering. One use of the phrase in the early eighteenth century was "on laissez faire la nature," or "let nature run its course,"[3] a prescription that only makes sense if nature writ large is harmonious and self-regulating. Ecologists have largely abandoned the idea that nature left undisturbed achieves a harmonious balance.[4] Instead, ecological systems are frequently out of equilibrium and can settle into many "basins of attraction" (locally stable configurations) that vary in their desirability from a human perspective. Wise ecological policy requires active management with complex systems in mind. The same is true for wise economic policy, but the pre-Darwinian concept of laissez-faire gets in the way of seeing this clearly.[5] I will have much more to say about economic policy as we proceed. For now, the important point to make is that we still make massive errors in functional reasoning.

Darwin's theory of evolution was a breakthrough in how to think properly about the presence and absence of functional design in the natural world. Virtually every trait that appears well designed in nature—the wing of the bird, the fang of the tiger, the speed of the gazelle—exists due to a historical process whereby individuals possessing those traits survived and reproduced better than individuals possessing different traits (Tinbergen's history question). In evolutionary parlance, they were more fit. *Enhanced fitness compared to alternative traits is the only source of design in nature.*

This simple statement leads to one of the most powerful tools in Darwin's toolkit, which is often called "adaptationist thinking" or "natural selection thinking." If you want to understand the features of an organism in the same way as the features of an avocado cuber, ask how the organism would need to be structured to survive and reproduce in its environment (Tinbergen's function question). That's how I reasoned about the feeding behavior of copepods as a graduate student in the 1970s—my first introduction to natural selection thinking.

To say that adaptationist thinking is a fruitful line of inquiry does not mean that it is simple or leads directly to the correct interpretation of functional design. It also doesn't mean that every detail of an organism is adaptive.[6] Consider the red blood cells that are coursing through your veins. Their primary function is to carry oxygen to your other cells. They also happen to be colored red. Is their color part of their functional design? The question is not totally farfetched. Perhaps the color of blood has signal value in a social species such as our own. When we bleed, it is highly visible so that others can come to our aid. But if that were the case, then the color of blood would probably be different in species where individuals provide aid to each other, compared to less cooperative species. Since blood is red in both cooperative and non-cooperative vertebrate species, the social signaling hypothesis for its color is unlikely to be correct. The most judicious conclusion is that the color of blood has no function. It is just a by-product of the hemoglobin molecule's main job of carrying oxygen. To advance further down this line of inquiry, it would be helpful to know more about hemoglobin as a physical molecule, which requires pivoting to Tinbergen's mechanism question. Evolutionists form, test, and accept or reject hypotheses like this all the time as part of our job of arriving at the facts. It's part of our conceptual toolkit.

As another example, consider the behavior of birds on remote oceanic islands. When Darwin visited the Galápagos Islands, he was surprised when birds alighted on his shoulders and arms as if he were a tree. They were well adapted to island life in the absence of people, but poorly adapted to the presence of people and other mammals such as mice, rats, goats, pigs, cats, and dogs. The moment that people and their mammalian companions set foot on remote oceanic islands, the environment changed radically for all of the native inhabitants and what was adaptive in their old environment became tragically maladaptive in their new one. Unless the native fauna and flora are protected by people or adapt to their new environment, they are likely to go extinct. Adaptations must be understood in terms of survival and reproduction in the *historical* environments that gave rise to the adaptations (Tinbergen's history question). When the environment changes, there is no reason to expect organisms to be well adapted to their new circumstances, and generations are required for new adaptations to evolve.[7]

Evolutionary mismatches don't just happen to other species. In many respects, *we* are like those bird species on oceanic islands, adapted to environments that have ceased to exist and sadly mal-adapted to our current environments. Three detailed examples will be provided in the next chapter. For now, I hope you are beginning to see how Tinbergen's function and history questions—two important tools in the evolutionist's toolkit—might also come in handy as tools in a policymaker's toolkit.

LIVING SNOWFLAKES

If we want to understand the nature of a snowflake, there is only a physical process to consider. A snowflake is born when water molecules crystallize around a tiny airborne particle such as dust or pollen in freezing temperatures. Water molecules are shaped in a way that creates a hexagon when they join together to form a crystal. As the crystal grows, new molecules are added unevenly to the outer edge based on irregularities, creating arms that themselves provide edges for continuing growth. Since a typical snowflake includes an estimated 10^{19} water molecules (a mind-boggling number), every snowflake is indeed unique, as many of us are told as children. The fact that crystalline growth depends upon temperature and other weather conditions, with each snowflake experiencing different conditions as it falls toward earth, contributes to its uniqueness.

Understanding the formation and growth of a snowflake is an example of thinking in physical reasoning mode, based entirely on the interaction between the chemical properties of water and its physical environment. Functional reasoning is not needed and indeed would be misleading, since a snowflake is neither a human artifact nor a product of evolution.

For life forms and human artifacts, it is necessary to think in *both* physical and functional reasoning modes, which is why Tinbergen's mechanism and development questions are needed to complement his function and history questions. Like a snowflake, this beautiful image of an anglefish skeleton is the product of a physical process that grows and develops over a period of time. Unlike a snowflake, living physical processes have been shaped by natural selection over the course of billions of years to replicate with astonishingly high fidelity—and the more one learns about the details, the more wondrous they become. The mechanistic branches of the biological sciences, such as biochemistry, molecular biology, genetics, and neurobiology, are devoted to revealing the physical basis of life.

It is remarkable how the function of an object and its physical basis can be studied independently of each other. Suppose I challenged you to make an avocado cuber out of physical materials that are completely different from the one shown in the illustration. You could probably meet my challenge without much difficulty. Likewise, consider the mundane fact that many desert-living species are sandy colored to conceal themselves from their predators and prey. This observation holds for desert-living snails, insects, amphibians, reptiles, birds, and mammals. The reason in each case

Unlike a snowflake, this angelfish skeleton has both beautiful structure and function.

is clear enough—individuals varied in their coloration and those that matched their background the best survived and reproduced the most. But wait! The exteriors of these species are made of completely different physical materials—calcium carbonate for snails, chitin for insects, keratin for reptiles, and so on. We didn't need to know anything about the physical makeup of each species to predict that they should be sandy colored. As long as the physical makeup of each species results in heritable variation, then it becomes a kind of malleable clay to be shaped by selection.[8]

It's an exaggeration to say that species can evolve in any direction. There are constraints imposed by their genes and developmental pathways. Hence, some features of organisms can only be understood in terms of their physical makeup and not their functional design. Also, full understanding of any particular adaptation requires mechanistic understanding. We might not need to know that the shells of desert snails are made from calcium carbonate to predict that they should be sandy colored, but we will need to know that fact to understand how it happens, how it sometimes goes wrong, and so on.

A MASTER EVOLUTIONIST AT WORK

The best way to learn a craft is by watching a master craftsperson at work. I will therefore close my introduction to Darwin's toolkit by showing how one master evolutionist combines Tinbergen's four questions to study evolution in action.[9] Richard Lenski, the son of the sociologist Gerhard Lenski and the poet Jean Lenski, was born in 1956. His interest in ecology and evolution caused him to specialize in the study of insects in their natural environments. Shortly after he received his PhD from the University of North Carolina in 1982, however, Lenski had a change of heart. If he really wanted to study fundamental issues in ecology and evolution, he should be studying a microbe named *Escherichia coli*.

Why? Primarily because *E. coli*, a species of bacteria that inhabits our guts, reproduces every 20 minutes under optimum growth conditions. That works out to 72 generations every day and 504

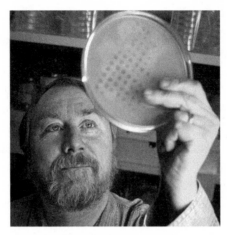

Richard Lenski: A master craftsman
at work

generations every week. If we pick a rough figure of 20 years for a
human generation, then a week's worth of evolution for *E. coli* (at
its maximum growth rate) is the equivalent of over 10,000 years of
evolution for us, which would take us back to the dawn of human
civilization.

Also, *E. coli* is one of the best-studied organisms in the world.
Biologists have wisely selected a few species that they call model
organisms, based on the strategy that it is better to know a whole
lot about a few organisms than a little about a lot of organisms. The
A-list of model organisms includes the bacteria *E. coli*, the nematode
worm *Caenorhabditis elegans*, the fruit fly *Drosophila melanogaster*,
the mouse *Mus musculus*, and the rat *Rattus norvegicus*. It is difficult
to fathom the degree to which these model organisms have been
examined from stem to stern by biologists, including a complete
inventory of their genes and extensive knowledge of how the genes
interact to assemble the whole organism (Tinbergen's mechanism
and development questions).

A microbe such as *E. coli* has another advantage that is truly
extraordinary for evolutionary research. The cells can be frozen for
extended periods and brought back to life by thawing them. This
meant that Lenski could place *E. coli* in a novel laboratory environ-

ment, periodically freeze samples, and bring the ancestors back to life to compare with their own descendants. A living fossil record!

The experiment that Lenski started in 1988 was simplicity itself. A single genetically identical clone of *E. coli* was used to create 12 populations inhabiting conical glass beakers called Erlenmeyer flasks in the laboratory. Each flask contained 10 ml of liquid growth medium with glucose as the main energy source. The flasks were placed on a shaking incubator that kept the *E. coli* well mixed under constant temperature and light conditions. Every day without fail, 0.1 ml of medium from each flask was transferred to a fresh flask with 9.9 ml of fresh sterilized medium. The *E. coli* from the 12 populations were never mixed. Even though the flasks were arranged in neat rows and columns on the shaking incubator, they were as genetically isolated from each other as if they were oceanic islands thousands of miles apart.

The *E. coli* grew well in their new laboratory environment, firing off about seven generations every day before depleting their resources. Every 500 generations, a sample from the flasks was frozen to become part of the living fossil record. Although the procedure sounds simple, it required monklike devotion on the part of Lenski and his associates. The transfers had to take place every day without fail. Someone always had to be on duty. Backup flasks and frozen samples had to be maintained. Periodic checks for contamination were required. Provisions for electrical outages and equipment failures had to be made. Day after day after day, right up to the present (nearly 70,000 generations as of this writing). In human terms, that is the equivalent of our history as a species!

In each of the 12 flasks, genetic mutations arose that caused some *E. coli* to grow faster than others. These strains increased in frequency until they replaced the slower-growing strains. Lenski knows this with certainty because he can compare the growth rates of the descendants with their own ancestors by unfreezing the earlier samples. When they are grown side by side in a horse race, the descendants grow 60 percent faster than their ancestors (Tinbergen's function and history questions).

Using the techniques of molecular biology, Lenski can also pinpoint the genetic changes that took place and how the new genes

altered the anatomy and physiology of *E. coli* to adapt it to its laboratory environment (Tinbergen's mechanism and development
questions). Do you think that the genetic changes were the same
or different in the 12 populations? The answer to this question is
not obvious. On one hand, each population inhabited exactly the
same chemical and physical environment, so whatever anatomical
and physiological changes work for one population should work for
the others. On the other hand, genetic mutations are chance events
and it is unlikely that the exact same mutation would occur in the
different populations (Tinbergen's history question).

This combination of functional similarities (all strains good at
digesting glucose) and genetic differences among the 12 populations
is exactly what Lenski observed. The body size of *E. coli* increased
in all 12 populations, for example, but the particular mutations that
caused body size to increase in a mechanistic sense differed. There
are many ways to skin a cat, and there are many genetic changes in
E. coli that are functionally equivalent as far as increasing body size
and other adaptations are concerned.

After the 12 populations had been isolated for 2,000 generations, Lenski started a new experiment while keeping the old experiment going. He switched the energy source for the 12 populations
from glucose to another sugar called maltose. What do you think
happened? All 12 populations were derived from a single clone and
had adapted to the exact same laboratory environment with glucose
as the energy source. Nevertheless, the populations varied greatly
in how well they could grow on maltose (Tinbergen's function
question). They had genetically adapted to glucose as an energy
source in different ways and these differences had consequences for
how well they could grow on maltose. All of the populations were
able to improve their growth rates on a maltose diet over the course
of 1,000 generations, but their starting points were very different,
unlike the start of the original experiment. Each population had
become a distinctive genetic entity (Tinbergen's history question).

Returning to the original experiment, you might think that the
12 populations would run out of ways of adapting to their unchang-
ing laboratory environment. It's true that they adapted more
quickly at the beginning of the experiment, but they never reached

a plateau. Even after many thousands of generations, new beneficial mutations arose to improve growth rate still further. These mutations might well have occurred during previous generations but failed to spread for lack of other genetic changes. In other words, as the 12 populations diverged, they provided different genetic backgrounds for subsequent mutations. The same mutation could be beneficial in one population and neutral or deleterious in other populations (Tinbergen's mechanism and development questions).

After Lenski's experiment had been in progress for fifteen years, something amazing happened. One population evolved the ability to digest a compound called citrate that had been present all along as part of the recipe of the growth medium. Citrate serves as an energy source for some bacteria but not for *E. coli*. For *E. coli* to start digesting citrate would be like for us to start digesting hay. Nevertheless, that is what one population started to do. Was this due to a once-in-a-trillion mutation that could have occurred in any of the 12 populations? Or had the genetic changes in one population during previous generations uniquely prepared it to adapt to an entirely new resource? Lenski and his associates were able to answer this question by reaching back into the frozen fossil record to a generation that preceded the evolution of the citrate-digesting mutation and restarting the experiment from that point. Sure enough, the same population that evolved the ability to digest citrate the first time did so again. Its unique genetic odyssey enabled it to do something that no population of *E. coli* had ever done before (Tinbergen's history question).

Science often requires the use of physical tools. For a microbiologist, these include Erlenmeyer flasks, petri dishes, autoclaves, and the like, which came into use over a century ago. To these we can add the vastly more sophisticated tools of molecular biology that make it possible to read the genetic code letter by letter, surgically snip out genes and insert them into other genomes, and perform other wizardry. Training is required to master these tools, but it is a remarkable fact that Lenski didn't start learning them until *after* he received his PhD. Retooling to become a microbiologist was little different from a carpenter retooling to become a plumber. Work is required but any motivated person can do it.

Far more important than the physical tools are the *conceptual* tools that Lenski used to design his experiments and interpret his results. First and foremost, he used adaptationist thinking to predict that *E. coli* would increase its growth rate in the laboratory environment that he created for them (Tinbergen's function question). He could make this prediction without knowing anything about the physical makeup of *E. coli* because he could confidently assume heritable variation in growth rate. Since evolution is a historical process that relies upon chance mutations, he predicted that it would not take exactly the same course in each population and that the populations would increasingly diverge over time (Tinbergen's history question). Studying the genetic changes that took place in each population required switching from functional reasoning mode to physical reasoning mode. Genes are physical entities and the symphony of genetic interactions that result in the development of an *E. coli* is a purely physical process, like the growth of a snowflake (Tinbergen's mechanism and development questions).

Function. History. Mechanism. Development. These are the conceptual tools that Lenski learned as part of his PhD training in evolutionary biology, which he used to inform his research on insects before switching to *E. coli*. They are the conceptual tools that biologists use to study all aspects of all species, as Niko Tinbergen wisely noted. Once mastered they become second nature, like knowing how to ride a bicycle. They might seem so simple that they couldn't be new—but only for those who have adopted evolution as their worldview. Now let's see how the evolutionary toolkit can become part of the policymaker's toolkit to address the problems of modern existence that everyone earnestly wants to solve.

3

Policy as a Branch of Biology

The challenge for Niko Tinbergen, Konrad Lorenz, and Karl von Frisch during the middle of the twentieth century was to show that ethology is a branch of biology—in other words, that behaviors evolve in much the same way as other traits. By the time they were awarded the Nobel Prize in Medicine in 1973, their mission had largely been accomplished—not only by their own efforts, but by many others employing a fully rounded four-question approach. Moreover, the historically separate fields of evolution, ecology, and behavior were becoming fused into a single discipline, often labeled by the acronym EEB, which was part of my training as a graduate student.

The challenge of this book is to show that policy is a branch of biology. A standard definition of policy is "a course or principle of action adopted or proposed by a government, group, or individual." Liberal and conservative politicians propose different policies to improve the economy. Many religions encourage the policy of "do unto others," at least in some situations. A "tiger mom" might adopt a policy of strict discipline toward her children. To view policy as a branch of biology means that our proposed actions must be deeply informed by evolution. Around the world, we should be consulting evolutionary theory at least as much as we consult our constitutions, political ideologies, sacred texts, and personal philosophies.

This is a radical claim, insofar as an evolutionary worldview is currently absent from virtually the entire policymaking universe. Nevertheless, I hope that it will make sense to you by the end of this book, in the same way that the claim of Tinbergen & Company made sense to the scientific community. In this chapter we will proceed on our journey with three stories that I have arranged in this order for a purpose. The first story describes why so many of us must wear glasses or contact lenses to see clearly. The second story describes how our immune system, an adaptation with a 500-million-year-old history within the vertebrate lineage, can go haywire in modern environments. The third story is about child development. All three stories highlight the importance of Tinbergen's development question and the concept of evolutionary mismatch, whereby adaptations to past environments can tragically misfire in current environments.

AS PLAIN AS THE GLASSES ON YOUR FACE

Our eyes are a perfect example of a functionally designed object. They capture images of the world in much the same way as a camera, complete with a lens, a photo-sensitive surface, and mechanisms for adjusting the amount of light and focal distance.

Nobody doubts that eyes are designed for seeing. Since eyes were not designed by people, they signify either the existence of another designing agent, such as a god, or a designing process, such as natural selection. That debate was settled long ago within the scientific community. In fact, what we know about eyes today is like a grand version of Richard Lenski's experiment on *E. coli*. Just as each of his twelve populations responded independently to the same selection pressure (the ability to use glucose as an energy source), over a hundred animal lineages responded independently to the selection pressure of using light as a source of information.[1] One of the most recent discoveries is a single-celled marine organism called *Warnowia*. While many single-celled organisms have evolved eyespots for orienting toward light, *Warnowia* has evolved a full-fledged eye with a cornea, lens, and retinal body constructed from subcellular components.[2]

Just as there are a limited number of ways that glucose can be processed as an energy source by *E. coli*, there are a limited number of ways that light can be processed as information—but the particular mechanism employed by any particular animal lineage is largely a matter of chance and its previous history. A famous example is the eye of the octopus and other cephalopods, which bears an uncanny resemblance to the vertebrate eye. A closer look reveals differences, however. The octopus eye develops from skin tissue while the vertebrate eye develops from brain tissue. Light strikes the retinal cells of octopus eyes directly but must pass through a layer of nerves and blood vessels to strike the retinal cells of vertebrates. Vertebrate and cephalopod eyes are *different* because they evolved independently from each other (Tinbergen's history question), and they are *similar* because they must have certain anatomical structures to function as eyes (Tinbergen's function question). All examples of convergent evolution share this combination of similarities and differences.

When we focus on Tinbergen's mechanism and development questions, eyes are created by a purely physical process, like the formation of a snowflake. A physical process that has not been shaped by selection tends to be variable in its outcome. Snowflakes

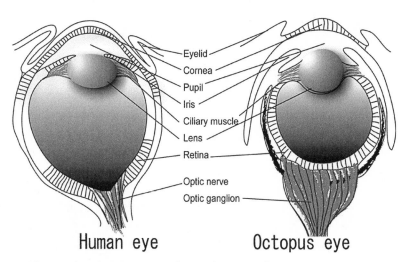

Eyelid
Cornea
Pupil
Iris
Ciliary muscle
Lens
Retina
Optic nerve
Optic ganglion

Human eye   Octopus eye

All examples of convergent evolution share a combination of similarities and differences.

are an extreme example in which every snowflake's encounter with its environment turns it into a unique individual. That kind of variability would be disastrous for eyes. Eons of natural selection have shaped the physical process of eye development to result in the same complex organ time after time. How is this miracle of quality control accomplished?[3]

Occasionally, abnormalities occur that provide a glimpse of the highly orchestrated process of normal eye development. Cataracts, for example, develop when the lens of the eye loses its transparency. This condition usually appears in older people but occasionally appears in newborn babies. When surgical procedures to implant artificial lenses were first developed, doctors were reluctant to perform the procedure on very young children, electing instead to wait until they were older. The results of this well-meaning decision were tragic. The patients remained blind or profoundly visually impaired, even though the original cause of their blindness had been corrected.

What happened to produce this tragic result? The doctors unknowingly assumed that the anatomical structures of the eye and the neural processing of visual information in the brain develop without any input from the environment. What they discovered the hard way is that normal development requires interaction with the external environment in the form of images striking the retina. A cataract prevents this from taking place. Without the environmental input, the physical process of eye and brain development cannot properly occur. The derailed developmental process cannot be restored by introducing the environmental input later in life.

This is a great example of the need to base policy (when to perform surgery on babies with cataracts) on biology (a knowledge of normal eye development). Doctors thought that they were doing the right thing by waiting to remove the cataracts, but their ignorance of the biology of eye development caused them to make a tragic mistake. As they learned more about how the eye develops and its need for environmental input, doctors were able to implement a wiser medical policy about when to perform cataract surgery on very young children.

The development of the eye and visual processing in the brain illustrate a principle that I call *rigid flexibility*. You already know about this principle if you have used computer software to help you prepare your taxes. The software prompts you for just the right kind of information, which it processes in just the right way to calculate what you owe or (hopefully!) should be refunded. Its ability to calculate anyone's taxes is impressively flexible, but it is useless for anything else and even fails to correctly prepare your taxes if you feed it the wrong information or change a few lines of code. Its flexibility is therefore rigid and easily broken. The concept of rigid flexibility sounds like a contradiction in terms but makes perfect sense from the right perspective.

The development of the eye and visual processing of the brain is rigidly flexible. Just as the software requires you to enter the right information, the developing eye requires the right environmental input to properly develop. Eons of selection have shaped the developmental process to require inputs that are reliably present in the environment, which is why eye development results in a functioning organ time after time. But change a few lines of the genetic code (such as a deleterious mutation) or create an artificial environment lacking the right inputs, and the process of eye development crashes.

As an example, the neural processing of visual information requires the recognition of contours. This is accomplished mechanistically by some nerve cells firing at the sight of horizontal lines, others firing at the sight of vertical lines, and others firing at the sight of oblique lines. The propensity of each type of nerve cell is hardwired at birth, but unless the cells actually fire in the infant organism, they fail to form the right connections with other nerve cells to process the information. Normally this is not a problem because all natural environments contain an abundance of horizontal, vertical, and oblique lines. What would happen if you raised an infant in a visual environment with only horizontal lines or only vertical lines? Experiments were performed on kittens in the mid-twentieth century to answer this question and resulted in profound visual impairment.

This apparatus interferes with eye development
without ever touching the eye.

Remarkably, the environments that people build for themselves
can hinder eye development, not by limiting the range of contours (as in the kitten experiments) but in other ways. People who
grow up in these abnormal human-built environments are unable
to focus on distant objects, a condition known as myopia or nearsightedness. Anatomically, myopia is caused by distortions in the
physical dimensions of the eyeball, the thickness and curvature of
the cornea, and the shape of the lens. Luckily, the condition can be
corrected with glasses or contact lenses. If it weren't for these cul-

tural inventions, myopia would be catastrophically debilitating, as any nearsighted person who has lost their glasses can attest!

The fact that myopia is caused in large part by an aberrant environment has been known for a long time. One study conducted in 1975 examined Inuits from two settlements in northern Canada.[4] Myopia was far more common in younger people than their elders, was more common in women than in men, and correlated with the amount of schooling in both sexes. The transition from hunter-gatherer life to a settled life was causing an epidemic of nearsightedness and schooling evidently had something to do with it.

A more recent study of Jewish teenagers documented that over 80 percent of boys who attended Orthodox schools were nearsighted, compared to around 30 percent of Orthodox girls and non-Orthodox Jews of both sexes.[5] The Orthodox boys spent as many as sixteen hours per day in their schools. National comparisons conducted between 1998 and 2004 show that the prevalence of myopia was less than 3 percent in Nepal (with very little schooling) and over 69 percent in Guangzhou province of China (with a *lot* of schooling).

Clearly, modern human-built environments and/or modern human activities (such as a large amount of time spent reading) disrupts the normal process of eye development. One possibility is the amount of time spent focusing on close objects required by reading or certain occupations such as checking for weaving faults in the textile industry, where the prevalence of late-onset myopia is exceptionally high. However, there is mounting evidence that the main environmental culprit is not the amount of time spent focusing on close objects but the amount of time spent indoors.

In one study that capitalized on a natural experiment, ethnically Chinese children (ages 6–7) living in Singapore were compared to ethnically Chinese children living in Australia.[6] There was a whopping ninefold difference in the prevalence of myopia, and the main environmental difference between the two groups was not the amount of time doing schoolwork but the amount of time spent outdoors—an average of fourteen hours a week for the kids in Australia versus an average of three hours a week in Singapore. A 2012 review of all relevant studies to date estimated a 2 percent reduced odds of myopia for each additional hour of time spent outdoors per week.

We are still in the process of improving our theories on the environmental causes of myopia. The theory that myopia is caused by spending large amounts of time focusing on close objects leads to one set of policy recommendations. The theory that myopia is caused by time spent indoors leads to another set of policy recommendations. One theory might be wrong or both might be correct to a degree, and the causes that they postulate might interact in a complex fashion. Only more scientific inquiry can decide the facts of the matter—and only the facts of the matter should inform public policy. Hence, the policy must be based on a detailed understanding of biology. Perhaps this seems obvious to you for something as "biological" as eye development, but my second story will expand the boundary of what is typically considered "biological" into what is typically considered "behavioral" and "social."

CLEANLINESS IS NOT NEXT TO GODLINESS

In 1847, an obstetrician named Ignaz Semmelweis noticed that pregnant women examined by doctors at the Vienna General Hospital were much more likely to die from puerperal fever than women attended by midwives. More sleuthing revealed that the examining doctors usually came directly from performing autopsies. By the simple expedient of making the doctors wash their hands with chlorinated limewater before examining pregnant women, Semmelweis was able to reduce the mortality rate from 18 percent to 2.2 percent.[7]

This was a milestone in the acceptance of the germ theory of disease, which had been proposed in various forms since the sixteenth century but wasn't widely accepted until the end of the nineteenth century.[8] It's easy to conclude from germ theory that the best policy is to get rid of all of them. "Kills 99.9% of all germs" is a frequent boast of disinfectant products. But this reasonable assumption, based on previous scientific advances, turns out to be wrong. Environments that are too clean can lead to diseases of their own.

How can this be? One reason is that not all of the creatures that reside on and inside us are harmful. Many of them, in fact, are helpful. Another reason is more subtle: Regardless of whether the creatures are harmful, helpful, or neutral, *they have always been there*. Removing them creates an environment that never before existed in the history of our species or the species that gave rise to *Homo sapiens*. We have just seen what happens to the development of our visual system in abnormal environments. Something similar happens to the development of our immune systems in overly hygienic environments, and the problems of crashing the immune system can be far more difficult to solve than myopia.

The fact that we serve as a habitat for a teeming ecosystem of microbes and other tiny creatures has only recently become a focus among biologists. A term for the ecosystem, microbiome, was coined around the turn of the twenty-first century.[9] Mind-boggling numbers are required to describe the size and diversity of our microbiomes. We each begin as a single cell, the union of a sperm and egg, which divides again and again to create and maintain our bodies. As adults, we are made up of approximately 39 trillion cells that interact in a symphony of cooperation to keep us alive. The organisms that use us as a habitat, as if we were a miniature planet, are *on the same order of magnitude as our own cells*.[10] A bacterial cell is much smaller than our cells, so the combined weight of our microbiome is a small fraction of our weight. Our microbiomes include bacteria, viruses, protozoa, and multicellular organisms such as worms and mites. A wild guess for the number of species inhabiting you at this moment is 10,000, but they have never been counted and most of them have never been cultured outside the human body. In addition, the whole concept of a species breaks down for bacteria because of all the gene swapping that takes place. When we bypass the species level and count all the different genes in our microbiomes, they outnumber our own genes by a factor of about 200 to 1.

If you are accustomed to thinking that "cleanliness is next to godliness," here is a parable to help you get over the idea that all germs are bad and need to be washed away. Imagine that you are a farmer relaxing after a long day. Farming is hard work but it is also immensely satisfying. The horses and cows are in their stables, the pigs in their pen, and the chickens in their coop. There will be eggs, bacon, and fresh milk for breakfast. The garden is growing taller every day. The weeds have been removed, the corn is in neat rows, and the tomatoes are just starting to ripen. Your dog, who has been padding along at your side all day, is curled up at your feet. She'll start barking if her sensitive ears detect any noises outside, such as a fox approaching the chicken coop. Your house cat is on your lap and the barn cats, too numerous to count, are busy catching mice and having their own litters. Now imagine getting rid of all *these* species. You would be devastated. If you were a subsistence farmer, you would be dead within weeks. That's how you should think, at least in part, about your microbiome.

Notice that not all species belong on the farm. There are foxes, mice, weeds, and insect pests that need to be gotten rid of. Work is required to maintain the right species composition on the farm, and some of the species, such as your dog and cats, help with the work. Your microbiome is like a farm that needs to be well managed, and smart policy requires something in between "get rid of everything" and "do nothing."

Our immune system plays the role of the farmer in my parable, working hard to get rid of the species in our microbiota that don't belong. Like vision, the immune system operates beneath conscious awareness. We don't need to think about seeing or about managing our microbiota. In both cases, however, the physical processes that evolved by genetic evolution are mind-bogglingly complex and can only be understood by scientific inquiry. The immune system includes dozens of specialized cell types that work together to hunt down and weed out the pest species on the microbiotic farm. The cell types communicate at a distance using chemical signals, calling each other to infection sites. The immune system even includes its own evolutionary process that produces roughly a hundred million different antibodies and selects those that successfully bind to the surfaces of pathogens, tagging them for removal by other components of the system.

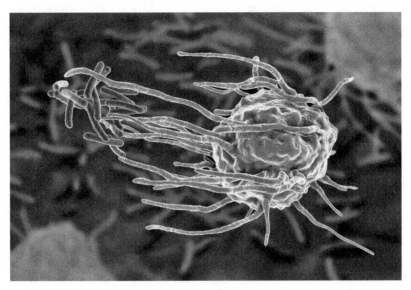

A macrophage weeding out species that don't belong on the microbiotic farm

The immune system must also develop in the fetus and during childhood. And just like the eye and visual processing in the brain, the immune system requires inputs from the environment for that to happen. It can't weed out every organism that it encounters. Somehow the immune system must learn to distinguish the friends from the foes, and the friends need to be present for the learning process to take place. When the friends are removed, then the immune system develops abnormally and can become like a farmer who attacks his own livestock, crops, and even himself.

The list of ailments that can result from a compromised immune system is long and includes anxiety disorders, asthma, autism, cardiovascular disease, depression, diabetes, eczema, hay fever, inflammatory bowel diseases, multiple sclerosis, and schizophrenia. Notice that some of these disorders are "physical" (e.g., inflammatory bowel disease) and others "behavioral" (e.g., depression). All disorders have physical causes, however, so the distinction is superficial. Indeed, the immune system is increasingly seen as having a behavioral dimension in addition to a physiological dimension.

Behavioral responses such as disgust help prevent us from contracting diseases in the first place. When we get sick, we exhibit a whole syndrome of behaviors: social withdrawal, decreased appetite, lethargy, and a lack of interest in usually pleasurable activities. These are adaptive responses to being sick, just as fleeing is an adaptive response to encountering a predator. They also largely overlap with the symptoms of depression. If the immune system includes both a physiological and a behavioral component, it makes sense that both might be affected when the immune system develops abnormally.[11]

Most of these disorders are more common in highly developed countries than in countries that live closer to a state of nature—just like the prevalence of myopia—which is one clue that a mismatch between modern environments and ancient developmental processes is the culprit. Other clues in an unfolding scientific detective story include the following:

- Children delivered by cesarean section have a higher frequency of allergic disorders than children delivered by normal birth. Evidently, passage through the birth canal inoculates the newborn with the microbiota of the mother.

- The use of antibiotics by women during pregnancy increases the frequency of allergic disorders in their children. In this case, even children who pass through the birth canal don't get the normal microbiota.

- Exposure to farm environments during pregnancy or the neonatal period decreases the frequency of allergic disorders in children.

- Exposure to natural environments up to the ages of 10–15 reduces the risk of multiple sclerosis later in life.

- Use of disinfectants in the home increases the prevalence of allergic disorders. Killing 99.9 percent of all germs is not a good thing!

- Restoring a more natural species composition in one's microbiome, including helminth worms, can result in immediate health benefits in adults.

- Chronic inflammatory disorders are more frequent in developed nations. And within these countries, they are more prevalent in urban environments than in rural ones. They include depression, schizophrenia, and autism, which many people do not realize involve an inflammatory component.

- People who emigrate from undeveloped nations to developed nations experience increases in inflammatory disorders. Some of the best studies are conducted on children from poor countries adopted by families from developed countries such as Sweden, the United States, and Israel. These children live in middle- and upper-class homes so their maladies cannot be attributed to poor living conditions in their adopted country.

- When dust is sampled from the rooms of children, there is a negative correlation between the diversity of microbes in the dust and the risk of asthma. The more diversity there is, the less asthma occurs.

- In one study that capitalized on a natural experiment, a genetically homogeneous human population straddles the national boundaries of Finland and Russia. The prevalence of type 1 diabetes is four times greater on the Finnish side than on the Russian side. This difference is accompanied by a striking difference in microbial diversity sampled from homes.

- A positive correlation between reduced gut microbial biodiversity and poor control of inflammation is a common finding in animal experiments.

- Institutionalized elderly people have a diminished microbiota, which correlates with poor health associated with
  inflammatory diseases.

These and other clues are summarized in a 2013 article by Graham A. Rook, a microbiologist who calls our microbiomes "old
friends." They have been with us for so long (eons) that we cannot live without them. In a 2015 radio interview he said: "We are
not individuals—it is a shocking fact for people to hear this—but
we are in fact ecosystems."[12] Most people think of ecosystems as
something that is *out there*, such as a forest or a lake. Many people
think that the most pristine and aesthetically pleasing ecosystems
are *without us*, such as wilderness areas. Advocates for the environment appreciate the need for stewardship of the earth's ecosystems,
all the way up to the planetary scale. Against this background, the
idea that each of us is a planet that requires environmental stewardship is indeed new for almost everyone. Moreover, your personal ecosystem is intimately connected to the wider ecosystem, as
a remarkable study headed by the Finnish evolutionary biologist
Ilkka Hanski shows.[13]

Hanski's team randomly selected a sample of 118 adolescents
living in a 100-by-150-kilometer area of Finland that is environmentally heterogeneous. First, each person was measured for an
inflammatory disorder called atopic sensitization, which involves
the propensity of the immune system to develop antibodies in
response to allergens. Second, the forearm of each person was
swabbed to obtain a sample of the microbial community on their
skin, which was identified to the level of genus using DNA methods. Third, the amount of vegetation cover in their yard was measured. Fourth, the major land use types within three kilometers of
their home was measured. Hanski's team was studying the *linkages*
between a human malady (atopic sensitization) and the ecosystem
inhabited by each person at three spatial scales—their skin, their
yard, and the larger area surrounding their home.

The skin microbiota for all 118 people combined included
572 bacterial genera in 43 classes. That's like a tropical rain forest!

Those who were sensitive to allergens had a low diversity of one kind of bacteria (called Gammaproteobacteria), and the statistical association was highly significant. None of the other kinds of bacteria were associated with atopic sensitization. Moving outward in scale, the diversity of the Gammaproteobacteria didn't have much to do with the vegetation cover of the yard, but it was strongly related to the amount of forest and agricultural land (as opposed to water and the built environment) within a three-kilometer radius. The skin ecosystem was clearly linked to the wider ecosystem.

Thus ends my second story about the need for policy to be informed by biology. It provides another demonstration that the theory decides what we can observe. The advent of germ theory during the nineteenth century enabled us to see the nature of infectious disease, leading to medical and public health practices that vastly improved the quality of our lives. However, a version of germ theory that treats all germs as bad is blind to another set of problems. The amount of suffering caused by the so-called "diseases of civilization" today rivals the amount of suffering caused by infectious diseases in the nineteenth century. A version of germ theory that takes microbiomes and the immune system into account promises to alleviate much of this suffering. As with myopia, some of the solutions might be extraordinarily simple, such as spending more time in natural environments or inoculating babies with a healthy microbiome. Other solutions will undoubtedly be more complex. Either way, the solutions can't be seen without the right theory, which is firmly rooted in biology.

My second story expanded the boundary of what is typically considered "biological" (such as asthma and diabetes) into what is typically considered "behavioral" (such as depression and anxiety). Once we appreciate that all behavioral traits have a physical mechanistic basis, no less than anatomical and physiological traits, the distinction between "biological" and "behavioral" disappears, as Tinbergen wisely noted so long ago. My third example expands the boundary still further—including the earnest desire of parents to do well by their children.

AN EGG WITH A VIEW

Many developmental processes are rigidly flexible. They evolved to receive just the right environmental inputs, which are processed in just the right way to lead to adaptive outcomes. But they can be subverted by environmental inputs that were not part of the "Environment of Evolutionary Adaptedness."[14]

One experiment that was performed on bird eggs provides an elegant (if perverse!) demonstration of rigid flexibility. Sound travels through the shell of a bird egg much better than light. As a result, the auditory system can start to receive environmental input while the developing chick is still in the egg, but the visual system must wait until the chick hatches. That's how it's been ever since birds evolved from dinosaurs. What happens if you accelerate the input of visual information by installing a window into a bird's egg? The installation is a snap to perform but it creates a completely different environment as far as bird development is concerned.

The window installation experiment was performed on bobwhite quail by the developmental biologist Robert Lickliter and colleagues in the 1980s.[15] Reading his articles is akin to watching a master scientist practicing his craft. Bobwhite quail are a commercially raised game species, so fertile eggs can be ordered from a sup-

Baby quail can hear their mothers before they can see them.

plier and raised in an incubator. Bird eggs have an air space at one end that enables the developing chick to start breathing and vocalizing before it hatches. The eggshell is porous enough to allow air to pass from the outside environment into the air space. The window installation involved removing the eggshell over the airspace on the twenty-first day of incubation, two days before hatching. In a preliminary experiment, Lickliter installed windows into eggs and placed the incubators in a dark room. These chicks developed in the same way as chicks from normal eggs, so the window installation by itself had no effect on their development.

In the main experiment, a 15-watt lightbulb was placed above the incubators and caused to pulse at 3 cycles per second, because the development of the visual system requires processing visual contrasts and not just a steady stream of light. Care was taken to ensure that the light did not alter temperature or humidity inside the incubator. The only difference between the experimental and control treatments was that the pulsing light reached the eyes of the chicks with windows but not those of the chicks inside normal eggs.

What happened? Bobwhite quail chicks are precocious and are capable of following their mother within a few hours after hatching. To follow their mother, it is critical for them to recognize her call. The auditory system develops while the chick is still in the egg. Installing a window causes the visual system to start developing prematurely, *and this interferes with the development of the auditory system*. Chicks that hatch from eggs with windows are unable to identify the calls of their mothers, which would be a catastrophic disability in their natural environment.

The fact that sensory systems develop in a certain sequence and at different rates turns out to be quite general. In birds and mammals alike, the order is tactile, vestibular (balance and spatial orientation), chemical (taste and smell), auditory, and visual. The sequence is governed in part by the availability of environmental inputs. If the visual system needs patterned light to develop, then it can't begin until after birth. In addition, there might be other reasons for the development of sensory systems to proceed in a certain order. Suppose that you are a contractor responsible for building a house with electricity, plumbing, and heating. You might have good reasons to install these systems in a certain sequence, even if they are eventually going to be integrated with each other. Whatever sequence you decide upon, an arbitrary change in the sequence is likely to be disruptive.

Against this background, consider a product called Bellybuds, a sound system for your unborn child. *You* like listening to Mozart, the Grateful Dead, or Snoop Dogg, so why deprive your unborn child of the same enjoyment? Or consider Baby Einstein, an entire line of educational products for babies that was started by a former teacher and stay-at-home mom and is now a division of Walt Disney. Might these seemingly innocent products be like installing a window into a bird's egg?

The earnest desire of most parents to do well by their children provides a fine example of theory determining what we can see. If we want our children to go to college and get good jobs, shouldn't we begin early? Shouldn't we buy them Baby Einstein products when they are still in their cribs and start teaching them the three Rs in preschool? All of this makes perfect sense when "practice makes perfect" is our guiding theory, just as killing all germs makes sense when we are guided by the wrong version of germ theory. But developmental processes are more complicated

If we want our kids to be well educated, shouldn't we start early? Maybe not.

than "practice makes perfect." As soon as we absorb this fact, then our entire perception of reality changes. That which previously made perfect sense and therefore guided our actions becomes dangerously misinformed.

Unlike bobwhite quail, humans are an altricial species, which means that a *lot* of development takes place after birth. Not just sensory development but emotional, social, sexual, and intellectual development. Given that we are such a behaviorally flexible species, in some respects development never ends. For thousands upon thousands of generations, human children were born into their societies and developed into functioning adults. An enormous amount of learned information was transmitted across generations. The social environments of our ancestors were diverse, just as visual environments are diverse (for instance, arctic vs. jungle). As with visual environments, however, ancestral human social environments shared enough in common for developmental processes to receive the appropriate environmental inputs and produce functional outcomes time after time. Now these inputs are in danger of being disrupted by modern social environments, including well-meaning but misguided efforts to do well by our children. It would be hard to imagine a more important priority for scientific research, but because the problem is largely unrecognized, scientific studies are few and far between. Here is some of the evidence that scientists are beginning to piece together:[16]

- Numerous studies have compared academically oriented preschool and kindergarten programs to programs that foster age-appropriate play. Academically oriented programs typically result in narrow academic gains over the short term that wash out within one to three years and in some cases are reversed. In other words, there is evidence that academically oriented programs for small children fail and might even backfire.

- There is evidence that academically oriented school programs for small children might harm social and emotional development over the long term. A study in Germany conducted during the 1970s compared the graduates of 50 play-based kindergartens with 50 academic, direct-instruction kindergartens.[17] Students who attended the academically oriented kindergartens showed short-term academic gains, but by grade 4 these students were performing significantly worse than students from play-based kindergartens. They were less advanced in reading and mathematics, and they hadn't adjusted as well socially and emotionally. Based in part on this study, Germany changed its policy and began favoring play-based kindergartens.

• One remarkable study led by David Weikart that started in 1967 followed a cohort of 68 children from high-poverty neighborhoods in Ypsilanti, Michigan, to the age of 23.[18] The children were assigned to one of three kinds of nursery school: Traditional (largely play-based), High/Scope (play-based with adult guidance), and Direct Instruction (an emphasis on reading, writing, and math using worksheets and tests). In addition, the families of the children were provided with guidance for home instruction similar to what their children were receiving in nursery school. As with other studies, the children receiving direct instruction showed early academic gains that subsequently disappeared. By age 15, there were no differences between the three groups of children in academic achievement but large differences in social and emotional development. Students in the direct instruction group had committed more than twice as many "acts of misconduct" as students in the other two groups. By age 23 the difference had become even more dramatic. Young adults that had received direct instruction as preschoolers were more emotionally impaired, less likely to be married and living with their spouse, and more likely to have committed a crime than young adults from the other two groups—39 percent (compared to 13.5 percent) had felony arrest records and 19 percent (compared to 0 percent) had been cited for assault.

- For children less than two years of age, two-dimensional images like books and screens are far less effective learning tools than three-dimensional objects (which include other people).[19] Two-dimensional images were a very small part of human experience for most of our evolutionary history. They have become omnipresent in modern life, so we clearly can and must develop the ability to comprehend them. But trying to accelerate this process might well interfere with other important developmental steps.

- Despite their popularity and promotion by commercial interests, there is no evidence that educational DVDs and videos for very young children are effective. One study conducted on 12- and 18-month-old children over a one-month period compared four treatment groups: 1) watching a word learning video with the parent encouraged to interact; 2) watching the same video without parent interaction; 3) interacting with the parent without the video; 4) a control condition with no instruction of any kind. Only the third group learned more words than the control group.[20]

- In addition to being ineffective, educational DVDs and videos for very small children can *retard* the very abilities that they are supposed to accelerate. In one study conducted in 2007, every hour of a word-learning DVD/video watched by 8- to 16-month-olds was associated with 6 to 8 *fewer* vocabulary words.[21]

- Evidence is accumulating that background media of all sorts (such as television) interfere with early child development, including executive functioning—the ability to plan and regulate behavior. The evidence is so clear that the American Academy of Pediatrics has recommended since 1999 that children younger than two should not be exposed to television and other background media. In one 2014 study led by Jenny S. Radesky at Boston Medical Center, higher levels of media exposure at nine months of age were associated with more irritability, distractibility, failure to delay gratification, and difficulty shifting focus from one task to another—even after controlling for other family and parental characteristics.[22]

- Good old-fashioned play, minimally supervised by adults, is emerging as important for the development of executive functioning. Play is where children learn how to regulate their behavior toward others in a safe and secure atmosphere. Children are highly motivated to play. No one needs to teach them how to do it. Yet, the opportunities for unstructured play are becoming extremely limited in modern life. And it's not just in schools, but also in overly scripted recreational activities, and in neighborhoods that are considered too unsafe for children to play without adult supervision.[23]

Thus ends my third story illustrating how policy should be considered a branch of biology. What does it add to the other examples? The first conforms to what most people associate with the word "biology." If the development of the eye and visual processing in the brain isn't biological, what would be? Nevertheless, this biological knowledge is necessary to make policy decisions on important matters such as when to remove cataracts in babies and what might be done about the epidemic of myopia in modern life. The skeptical reader, at the very least, must grant my point that policy should be based on biology for eye development.

My second story about the development of the vertebrate immune system also falls squarely into what most people would call biological, but some manifestations of immune system dysfunction are behavioral, such as anxiety, depression, and autism spectrum disorders. My second example therefore reinforces the point made by Tinbergen over a half century ago, that the study of behavior is a branch of biology. No matter how immune system dysfunction manifests, there is an urgent need to understand why it happens and to do something about it; in other words, to develop *policies* on the basis of *biological knowledge*.

My third example also starts out biological—the development of sensory abilities in birds—but ends up mostly behavioral. If how we raise our children at home and educate them in school isn't behavioral, what would be? Yet it is also profoundly biological, as I hope you now agree. The fact is that there is no meaningful distinction to be made between "biological" and "behavioral." An evolutionary worldview won't be fully established until the essential insight of Tinbergen becomes common sense for all of us in a wide variety of social policy situations.

I also sequenced my three examples to give a sense of urgency to the need to base policy on biological knowledge. It's easy to be complacent about the first example because the problem (myopia) has a relatively simple solution (corrective lenses) that has already been worked out. But immune system disorders haven't been worked out and while some of the solutions might be simple, others will surely be complex. With disruptions of child development, we are faced with the tragic possibility that we are harming our own children, based on our lack of biological knowledge.

Why are some people threatened by the idea of policy as a branch of biology? One threatening connotation is "inability to change." If something about us is biological, doesn't that mean that we are born a certain way and can't do anything about it? Another threatening connotation is the justification of inequality. If some people are born a certain way, might that be used to justify certain types of discrimination?

Once we get over the bogeyman story of social Darwinism, we can define the word "biology" the way that biologists do—the study of living processes—and begin to address policy issues using the biologist's conceptual toolkit, as admirably summarized by Tinbergen's four questions. Each of my stories involved an interplay of the four questions with development occupying center stage. Each story shows that development involves an interaction between an organism and its environment and that it is a highly scripted interaction written by an evolutionary process. Departing from the script can cause the process to crash. To the best of our current knowledge, if children spend most of their time indoors on close-focus activities, then they are likely to become myopic. If they live in overly hygienic environments, then they are likely to suffer from inflammatory disorders. If they don't engage in age-appropriate activities at home and in school, then they might be at risk for impaired executive functioning. That's threatening. On the other hand, there is something we can do about it by learning enough about such developmental processes to avoid these hazards. Basing policy on biology becomes an essential part of the solution, and that's alluring.

These stories also provide confirmation that theory decides what we can observe. In each case, harmful policies were practiced because they made sense against the background of certain ideas about the world. It made sense to wait before removing cataracts from babies, to rid the world of germs, and to start teaching our children the three Rs in their cribs. New ideas about the world were required to reveal why these practices might not make sense and to suggest other solutions that were previously invisible.

Better theories are only the beginning. No theory leads directly to the truth. The best that any theory can do is to outline a number of plausible hypotheses, which then need to be empirically tested. Armed with a theory of eye development that is approximately correct, we can say with confidence that myopia is caused by an environmental disruption of normal eye development. But is the disruption largely caused by factors such as spending too much time focusing at close distances, spending too much time indoors at low ambient light levels, and possibly other unknown behaviors? Only more scientific research can tell.

In the following chapters I will tell more stories that show how policy should be considered a branch of biology. We will see that no matter how far afield a given policy area seems to be from "biology" as often imagined by policy experts and the general public, all roads lead to an expanded conception of biology that can be studied by an interplay of Tinbergen's four questions.

4

The Problem of Goodness

The idea that harmony and order exist at large scales, such as human societies, the biosphere, and the physical universe, is deeply entrenched in Western thought and many other cultural traditions. It suffuses the Christian worldview and shows up in places that are not typically associated with Christianity, such as economics and complex systems science. In Christian thought, the belief that the universe was created by a benign and all-powerful god leads to a conundrum: If true, how can we explain the existence of evil? This is called the problem of evil and much ink has been spilled by theologians trying to resolve it—including Thomas Malthus's belief that famine and disease are divinely imposed to teach virtuous behavior.

The evolutionary worldview turns the problem of evil on its head. The behaviors that we associate with evil are easy to explain from an evolutionary perspective, because they typically benefit the evildoer at the expense of others. The problem is to explain how the behaviors that people associate with goodness, which typically benefit others and society as a whole, can evolve by a Darwinian process. Much ink also has been spilled by evolutionary thinkers to resolve the problem of goodness, but they have made more progress than theologians, for the simple reason that evolution provides a better theory of living processes than creationism. In other words, we are in a position to provide a scientific account of how the behaviors associated with goodness can triumph over the behaviors associated with evil—or vice versa—depending upon environmental conditions.

In this chapter I will tell three stories about the eternal contest between good and evil and how we can tilt the playing field in favor of goodness. The first two stories are purely biological, although they are highly relevant to human welfare. The third story gets to the heart of what human morality is and how it can be strengthened. Before we begin, I must say a little more about the progress that evolutionists have made resolving the problem of goodness.[1]

MULTILEVEL SELECTION

Take a moment to imagine a morally perfect individual. Don't worry too much about the precise definition of morality. Just trust your own intuition. What words would you use to describe him or her? Now take a moment to imagine that person's opposite. What words would you use to describe evil incarnate? In short, what is your version of Dr. Jekyll and Mr. Hyde?

Robert Louis
Stevenson's story
explores the eternal
conflict between good
and evil.

I have played this game with audiences around the world and all levels of expertise, from elementary school students to theologians and philosophy professors. The answers are so consistent that I list them in a slide that is part of my presentation, knowing it is what I am going to hear. Words used to describe the morally perfect individual include loving, honest, brave, giving, unselfish, loyal, and so on. Words used to describe the opposite include greedy, murderous, selfish, deceptive, manipulative, uncaring, and so on. Was I able to read your mind as well?

Darwin encountered a problem when he tried to explain how the traits associated with goodness can evolve by natural selection. In every case, they seem vulnerable to exploitation by the traits associated with evil. If natural selection favors individuals who survive and reproduce better than others, and if goodness involves helping others to survive and reproduce even at one's own expense, then how can goodness evolve?

While goodness posed a problem for Darwin, the solution was not far to seek. Social behaviors are almost invariably expressed in groups that are small compared to the total evolving population—a fish school, a bird flock, a pride of lions, or a human tribe. This means that an evolving population is not just a population of individuals but also a population of *groups*. If individuals vary in their propensity for good and evil, then this variation will exist at two levels: variation among individuals within groups, and variation among groups within the entire population. While goodness might be vulnerable to evil within any particular group, groups whose members are loving, honest, brave, etc., to each other will robustly outcompete groups whose members are greedy, murderous, selfish, etc., to each other. Here is one of the passages from Darwin's *Descent of Man* where he describes natural selection as a two-level process and relates it to human morality:

> It must not be forgotten that although a high standard of morality gives but a slight or no advantage to each individual man and his children over other men of the same tribe, yet that an increase in the number of well-endowed men and advancement in the standard of morality will certainly

give an immense advantage to one tribe over another. There can be no doubt that a tribe including many members who, from possessing in a high degree the spirit of patriotism, fidelity, obedience, courage, and sympathy, were always ready to aid one another, and to sacrifice themselves for the common good, would be victorious over most other tribes; and this would be natural selection. At all times throughout the world tribes have supplanted other tribes; and as morality is one important element in their success, the standard of morality and the number of well-endowed men will thus everywhere tend to rise and increase.

Darwin doesn't comment on the glaring fact that in his scenario, moral behaviors are confined to members of one's own tribe and are often directed against members of other tribes. Group selection doesn't *eliminate* immoral behaviors so much as it *elevates* them to the level of between-group interactions. Altruism within groups can become a form of collective selfishness toward other groups. Nevertheless, we need to explain how within-group morality can evolve before we attempt to tackle more expansive forms of morality.

In a 2007 review article that I wrote with the famed evolutionary biologist Edward O. Wilson, we summarized Darwin's theory of two-level selection this way:

Selfishness beats altruism within groups. Altruistic groups beat selfish groups. Everything else is commentary.[2]

While this solution to the problem of goodness is simple, the ramifications are far-reaching. In the first place, for within-group morality to evolve, between-group selection pressures must be stronger than within-group selection pressures. There is no guarantee that this will always be the case. When within-group selection is the stronger force, then evil triumphs over good. There is no warrant for a worldview that claims that nature writ large embodies goodness. Goodness only emerges when between-group selection pressures outweigh within-group selection pressures.

In the second place, nature is more complicated than two-level selection, or a hierarchy of only individuals within groups. Individuals are themselves groups of cells and genes. Single-species social groups such as fish schools, bird flocks, and lion prides exist in ecosystems composed of many species and a nested hierarchy of scales, ultimately making up the whole biosphere. In the human world we have genes, individuals, families, villages, cities, provinces, and nations, all nested within what Marshall McLuhan dubbed the Global Village. The tug-of-war between levels of selection that I described for individuals in groups exists for all levels. What's good for me can be bad for my family. What's good for my family can be bad for my clan—all the way up to what's good for my nation can be bad for the Global Village.

In short, two-level selection needs to be expanded into a theory of multilevel selection (MLS), from genes to the planet. The problem of goodness can potentially be resolved at any level, but the conditions for higher-level selection to prevail over lower-level selection become more challenging as we ascend the rungs of the multitier hierarchy, as we shall see with my three stories of the eternal contest between good and evil.

THE EVIL OF CANCER

Each of us consists of trillions of cells that are differentiated into hundreds of cell types. Every cell is descended from a parent cell, all the way back to the original sperm and egg. Each cell division requires copying the DNA, which consists of thousands of genes and roughly four billion base pairs—the nucleotide "letters" that make up the genetic "alphabet."

With a few exceptions, all of our cells have the same genes. The way that cells differentiate is by expressing some genes and silencing others. When differentiated cells divide, the patterns of gene expression must be copied in addition to copying the entire complement of genes. The inheritance of patterns of gene expression is called epigenetics.[3]

Some of our cells have long lifetimes. Many of the cells in your brain or in your ovaries (if you are a woman) were present when you were a child. Other cells have a lifetime of weeks or even days. If you are a man, for example, the sperm cells in your testes are no more than a few weeks old. Skin cells, liver cells, the cells that line our gut, and the cells that make up our immune system have especially high turnover rates. An estimated 500 billion cell divisions take place in our bodies every day!

Cells must divide and differentiate in just the right way to play their appointed role in a multicellular organism. They must express only some genes and not others. They must stop dividing when an organ is fully developed. Cells even routinely commit suicide (called "programmed cell death") when their absence benefits the organism more than their presence. This symphony of cooperation is produced by natural selection. Very simply, organisms whose cells work together for the common good survive and reproduce and their properties are inherited by their offspring more often than organisms whose cells fail to do their part. The result is a living snowflake: a purely physical process (addressed by Tinbergen's mechanism and development questions) that—unlike real snowflakes—results in a reliable working copy time after time.

The replication of genes and patterns of gene expression during every cell division are amazingly accurate, but not perfect. Imagine that every book had to be transcribed by hand from another book, which was true before the invention of the printing press. Even the most careful transcriber is likely to make a few errors, which will then be perpetuated in future copies. So it is with our cells. With billions of base pairs that need to be replicated, almost every new cell contains a few copying errors, or mutations.

Some mutations don't make any detectable difference in the performance of the cell or organism. Others impair the performance of the cell without harming the organism. In fact, one reason that programmed cell death evolved is to rid the body of mutated cells. Still other mutations cause the cell to grow inappropriately at the expense of neighboring cells and to evade programmed cell death. The tissues that form from these cells are called neoplasms.

Neoplasm cells are not good for the organism but they are perversely good for themselves. After all, natural selection favors any entity that survives and reproduces better than neighboring entities. Neoplasm cells fit that description. It doesn't matter that they might turn into malignant tumors that result in the death of the organism, and therefore their own death. Natural selection has no foresight. It is simply a physical process of replacement that takes place on the basis of differential survival and reproduction in the here and now, regardless of the long-term consequences. The only way that long-term consequences become salient at the cell level is through a process of natural selection operating at the level of multicellular organisms, which can be thought of as group selection among tribes of cells.

Neoplasms are much more common than malignant cancers. In fact, we are only in the process of discovering how common. Look at this image of an eyelid. It is not a new eye shadow. It is a pictorial of neoplasms on the eyelid of a person between fifty-five and seventy-three years of age. Each circle represents a mutated skin cell that grew at the expense of the normal skin cells to create a tiny patch of its own kind. This person did not have skin cancer, but rather a condition called drooping eyelid that is corrected by surgically removing some of the skin of the eyelid, affording an opportunity for a team of scientists led by Peter Campbell at the University of Cambridge and the Wellcome Trust Sanger Institute in Britain to conduct a genetic analysis of skin samples like this

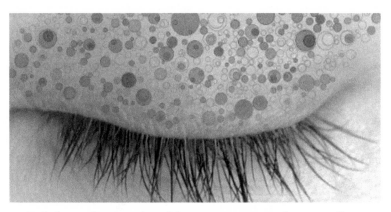

Each dot on this person's eyelid represents a group of mutant cells that has proliferated at the expense of the normal cells.

at a fine spatial scale.[4] They estimated that each skin cell had an average of about four mutations and over 20 percent of the cells had mutations associated with skin cancer. These are the cells that tended to grow at the expense of neighboring cells to form tiny neoplasms.

Each neoplasm is like one of the populations in Richard Lenski's *E. coli* experiment. Instead of being selected to digest glucose, they are selected to grow at the expense of neighboring cells. Additional mutations are required to turn a neoplasm into life-threatening cancer. The right combination of mutations to develop cancer only occurs in a small fraction of neoplasms. The rest are benign and do not threaten our health.

One key adaptation of a malignant cancer is to increase the rate of mutations. A fast-growing tumor isn't just one type of cancer cell that is rapidly proliferating. It is a boiling cauldron of hyper-mutating cell lines that compete against each other. Another adaptive strategy for a cancer cell is to disperse so that it can compete against normal cells rather than against other cancer cells—metastasis.

Cancer provides an example of multilevel selection and the eternal contest between the behaviors associated with good and evil. With cancer, the group is the multicellular organism and the individuals are the cells. Just as the traits associated with good are vulnerable to the traits associated with evil for organisms living in social groups, normal cells are vulnerable to cancer cells within multicellular organisms. In the same way groups of morally virtuous individuals outcompete groups crippled by selfishness, multicellular organisms free of cancer outcompete multicellular organisms riddled by cancer. The eternal struggle between good and evil takes place within our own bodies and has since the origin of multicellular organisms roughly a billion years ago.

Not only does cancer provide an excellent example of multilevel selection, but it also illustrates how the theory decides what we can observe. Cancer research is enormously sophisticated, but almost all of it is concentrated on Tinbergen's mechanism and development questions. The very concept of cancer as natural selection taking place within multicellular organisms wasn't proposed until the 1970s and still accounts for a tiny fraction of cancer research.[5] A fully rounded four-question approach, one that includes the function and history questions, challenges the wisdom of some current practices and suggests new possibilities for cancer treatments that were previously invisible.

Take the current practice of aggressive chemotherapy, which has the goal of entirely eliminating a tumor. This might make sense if all of the tumor cells are alike, but if they are a rapidly mutating and evolving population of cells, then aggressive chemotherapy becomes an extremely strong selection pressure for tumor cells that are resistant to chemotherapy. We already know that trying to blast bacterial diseases with antibiotics merely results in the selection of resistant strains that become ecologically dominant because other species in the ecosystem, such as predators and competitors, are completely wiped out. A more promising approach, informed by evolution and ecology, is to keep the ecosystem intact and enlist the aid of other species to help control the pest or disease species. A similar approach for cancer, led by the few researchers who are employing an evolutionary and ecological perspective, is called adaptive therapy.

Since the risk of cancer increases with every cell division, we might expect large and long-lived species such as elephants to be more prone to cancer than small and short-lived species such as mice. But this isn't the case—mice and elephants have roughly the same cancer rates. The reason appears to be that cancer is a greater selection pressure for elephants, so they have evolved more effective defenses against cancer than mice have. We should be studying large and long-lived species to learn how they suppress cancer so well and whether we can base therapies for humans on the same mechanisms. Cross-species comparisons, which rely heavily on Tinbergen's history question, provide a rich vein of information on cancer from an evolutionary perspective, but almost all cancer research is conducted on only two species, humans and laboratory mice—more blindness from the failure to employ the right theory.

Cancer from an evolutionary perspective is a fascinating and important topic in its own right and provides a great example of the contest between good and evil in unexpected places. Now let's see what chickens can teach us about the problem of goodness.

GOOD AND EVIL CHICKENS

In chapter 1 we met Francis Galton, Darwin's half cousin, who thought that people should be bred for their abilities in the same way that domesticated plants and animals are. Even Darwin thought that eugenics might work for humans, objecting only because he thought it would violate our instincts for sympathy and compassion, which he regarded as important human adaptations.

With this in mind, consider an experiment that was performed on chickens by William Muir and his colleagues at Purdue University's Department of Animal Sciences in the 1990s.[6] Their goal was to increase the egg-laying productivity of hens. Chickens evolved to live in flocks, but in the modern poultry industry they are often housed in cages with five to nine hens per cage, so the study was focused on maximizing production in this environment. The design was simple: they monitored how many eggs each hen laid, the most productive hens from each cage were used to breed the next generation, and so on for a number of generations. If the trait of egg productivity is heritable, then this method should substantially increase egg productivity over a number of generations, much in the same way that later generations of *E. coli* in Richard Lenski's experiment evolved to digest glucose more efficiently.

But that's not what happened. Instead, the subsequent generations laid *fewer* eggs and became more aggressive toward each other. The image opposite shows one of the cages after the experiment had been in progress for five generations. The cage originally housed nine hens, but six were murdered and the survivors had plucked each other's feathers. No wonder they weren't laying many eggs!

What happened to produce this ghoulish result? The most productive hens in each cage achieved their productivity by bullying the other hens. Bullying behavior is heritable in chickens, so selecting the biggest bullies led to a strain of hyper-aggressive hens within five generations. The energy expended and stress induced by their constant attacks on each other caused all of them to lay fewer eggs, despite the fact that the most productive had been selected to breed, generation after generation.

This is what happened when the "best" egg layer within each group was selected to breed the next generation.

In an experiment that was performed in parallel with the first, egg productivity was monitored at the level of whole cages. Instead of breeding the most productive hens within each cage, all of the hens from the most productive cages were selected to breed the next generation. The image on page 86 shows a cage from this experiment after five generations. All nine chickens are alive and fully feathered, and their egg productivity increased 160% during the course of the experiment.

These two experiments provide beautiful examples of within-group and between-group selection as envisioned by Darwin. The first one highlights the advantage that selfish traits have over cooperative traits within single groups. The chickens in the first image, tormenting and killing each other for their own gain, exhibit traits we would certainly call evil. The second experiment highlights the need for selection at the level of groups to evolve the traits that enable everyone within the group to thrive. The chickens in the second image, living amicably with each other, exhibit behaviors we would certainly call good.

Francis Galton assumed a simple relationship between individual abilities and societal warfare. Able societies are built by able

This is what happened when all hens from the best groups were selected to breed the next generation.

individuals. Ability is an individual trait that is inherited by offspring from their parents. Selecting the most able individuals therefore must result in the most able society.

But the chicken experiments suggest that this logic is flawed—even for farm animals where eugenics is a common practice. It seems Francis Galton was deeply mistaken about the relationship between individual abilities and societal welfare. The number of eggs laid by an individual hen is not an individual trait so much as it is a social trait, because it depends upon how members of the group behave toward each other. If the individuals who profit most from a social group do not contribute to the group's welfare, and if their traits are heritable, then selecting for them results in the collapse of the society. The relationships between single genes, observable traits measured in individuals, and the performance of whole groups are sufficiently complex—even for caged chickens—that selecting at lower levels often fails to produce the expected group-level outcomes. Selecting whole groups on the basis of their success proves to be more effective because the next generation will inherit some combination of all the individual traits that, in previous generations, resulted in fruitful interactions and contributed to the success of the group.

I will use the chicken experiments as a parable for human social interactions throughout the rest of this book, but for now let's examine the implications for agriculture and animal welfare, which are important policy areas in their own right. The most common commercial method of rearing hens is to cram them into standing-room-only cages, where they are unable to escape each other or perform their normal behaviors. Their beaks must be "trimmed" to prevent them from injuring each other, and their bones break from lack of use. If that's not injurious to animal welfare, what would be? Many people (including myself) are willing to pay more for eggs from hens that are allowed to range more freely, but free-range social environments have their own problems. Fighting still takes place and dominant birds prevent subordinate birds from accessing food, water, and nesting sites, resulting in low productivity for the group as a whole. Providing more space does not solve the problem of evil triumphing over good within groups, for chickens any more than ourselves!

A final lesson is that genetic evolution will take place in domesticated animals and plants, whether we want it to or not. It is sobering to contemplate the malleability of life. It only takes five generations to turn a population of mild-mannered chickens into a population of psychopaths. If we don't manage evolutionary processes, they will very likely take us where we don't want to go.

Thus ends my second story of the eternal contest between good and evil in unexpected places. Now it is time to bring the concepts to bear upon our own species.

WHAT IS MORALITY, ANYWAY?

At the beginning of this chapter, I wrote that there is no warrant for the deeply entrenched view that harmony and order exist at all scales of nature. We can expect harmony and order only when higher-level selection manages to prevail against lower-level selection. In animal societies, there are many cases of natural selection going the way of the first chicken experiment. The traits that we associate with evil triumph over the traits that we associate with good within groups, and the counterforce provided by between-group selection is not strong enough to save the day. These are "life's a bitch and then you die" societies. We would not want to live in them.

There are also cases where natural selection went the way of the second chicken experiment. Between-group selection is strong enough to prevail against within-group selection, favoring the traits that we associate with goodness. Many social species are mosaics of both kinds of traits, some maintained in the population by within-group selection, others by between-group selection. However, the balance between levels of selection is not static but can itself evolve. In rare cases, mechanisms evolve that largely suppress the potential for disruptive forms of selection within groups, making between-group selection the primary evolutionary force for most traits of the species. Then something magical happens: the group evolves to be so cooperative that it is transformed into a higher-level organism in its own right.

This transformation is called a major evolutionary transition, and it was first proposed in the 1970s by the cell biologist Lynn Margulis to explain how nucleated cells evolved from bacterial cells.[7] The former are much more complex, with internal structures such as mitochondria, chloroplasts, and ribosomes that are called "organelles" because of their specialized roles in maintaining the cell, similar to the organs of our body. Despite the difference in their complexity, it was assumed as a matter of course that nucleated cells must have evolved by small mutational steps from bacterial cells. Margulis's radical proposal was that nucleated cells originated as cooperative communities of bacterial cells. Decades were required for her symbiotic cell theory to be accepted, but now it is widely regarded as fact.

The idea that individuals can evolve from *groups* rather than from other *individuals* was then generalized in the 1990s by two theoretical biologists, John Maynard Smith and Eörs Szathmáry, to include the evolution of the first bacterial cells, the evolution of multicellular organisms, and the evolution of insect colonies.[8] In all cases, the higher-level entity evolves the properties of an organism by suppressing the potential for disruptive selection from within. Even the origin of life itself might be explained in this way as groups of cooperating molecular interactions.[9]

Multicellular organisms and their ability to suppress cancer provide some of the best examples of major evolutionary transitions, as we saw with my first story. Most genes that evolve in multicellular organisms do so by benefitting the whole organism relative to other organisms (or groups of organisms relative to other groups). It is comparatively rare for genes to evolve at the expense of other genes within the same organism, but only thanks to elaborative mechanisms that evolved to ensure that this does not happen. Higher-level selection is so much stronger than lower-level selection that we use a different word to describe the higher-level entity. Instead of calling it a society of cells, we call it an organism. This name change should not obscure the fact that a multicellular organism is nothing more than a highly regulated society of cells that evolved thanks to a very strong imbalance between levels of selection.

Social insect colonies such as ants, bees, wasps, and termites provide another outstanding example of major evolutionary transitions.[10] Unlike a multicellular organism, which has a clear physical boundary, the members of a social insect colony are physically separate from each other. On any given day, the honeybees from a single beehive can be dispersed over an area of several square kilometers. Nevertheless, their activities are so well coordinated that they invite comparison to a single multicellular organism. When searching for a new nest site, for example, a swarm of honeybees is every bit as discerning as a human house hunter, searching out and evaluating the alternatives on the basis of multiple criteria such as size, height, and exposure to the sun.[11] Words such as "eusocial," "ultrasocial," and "superorganism" are used to designate their adaptedness at the level of the colony, based on the fact that between-colony selection is the dominant evolutionary force.

People have been fascinated by the social insects since antiquity. Even though we are unlike them in so many ways, we feel an affinity for them and even treat their industry on behalf of their groups as an ideal for us to emulate. We also sometimes express an affinity and yearning for our societies to be like a multicellular organism, as in this passage from a seventeenth-century religious tract:

> True love means growth for the whole organism, whose members are interdependent and serve each other. That is the outward form of the inner working of the Spirit, the organism of the Body governed by Christ. We see the same thing among the bees, who all work with equal zeal gathering honey.[12]

These comparisons are metaphorical, but now they can be placed on a firm scientific foundation. *We are evolution's most recent major transition.* Almost everything that sets us apart from other primate species can be explained as forms of cooperation that evolved by between-group selection, thanks largely to our ability to suppress disruptive within-group selection. In most primate societies, group members are cooperative to a degree but are also riven by within-group conflict. Even the cooperation that exists often takes the form of coalitions warring with other coalitions within the same groups. To the best of our current knowledge, our distant ances-

Beehives have long been used as a
symbol of cooperation and industry.

tors evolved the ability to suppress bullying and other disruptive self-serving behaviors within groups, like multicellular organisms evolved ways to suppress cancer cells, so that the primary way to survive and reproduce was through teamwork.[13]

This brings us back to the topic of morality. At the beginning of this chapter, when I asked you to imagine the traits associated with good and evil, I encouraged you to trust your intuition and not to worry about formal definitions. Now we can do better. Few people are more authoritative on the subject than Simon Blackburn, who holds the Bertrand Russell Chair of Philosophy at Cambridge University in the UK. In an interview with Blackburn, I asked him to define morality as he would to students in a Philosophy 101 class, without reference to evolution. Here is what he said.

> At its simplest, morality is a system whereby we put pressure on ourselves and others to conform to certain kinds of behavior. That's the side of morality which is perhaps most obvious. It's associated with rules, with boundaries to conduct, and limits to criminal behavior when those rules are transgressed. On top of that, there is an element of morality that concerns our sentiments and emotions; for example our capacity to feel sympathy for each other's distress and a corresponding motivation to do something about it. So there are two sides to morality; one of them more gentle and humane and the other more coercive and to do with rules and social institutions designed to enforce those rules. For analytical purposes it is useful to separate them but in many contexts they merge into one another. For example, our sympathy for the distress of somebody being bullied might translate into a belief that the bully has transgressed social norms and the desire to punish them or to alleviate the distress in some way.[14]

This definition of morality, stated without reference to evolution, is exactly the system that we would expect to result from a major evolutionary transition. Our moral psychology is the societal equivalent of cancer-suppressing mechanisms in multicellular organisms. The coercive side of morality is required to suppress the potential for disruptive self-seeking behaviors within groups. Once the coercive side is established, then it becomes safe for group members to freely help each other without fear of exploitation.

This correspondence, between morality as already understood and morality as expected from an evolutionary perspective, was not lost on Blackburn. The rest of our interview explored the insights that can be gained from a more explicit study of human morality from an evolutionary perspective. No other theory comes close to explaining our odd mix of moral strengths and weaknesses: our intuitive sense of right and wrong; our virtuous behaviors and temptations to cheat; our zeal for monitoring and punishing the transgressions of others; and the ease with which we confine our virtuous behaviors to "us" and exclude "them." The more we see the problem of goodness through the lens of the right theory, the more we will be able to construct moral communities that are adapted to our modern age.

5

Evolution in Warp Drive

Darwin formulated his theory of natural selection in terms of variation, selection, and heredity, which is a resemblance between parents and offspring. At the time, the fact of heredity could be easily observed but the mechanisms of heredity were mysterious. That's why the work of Gregor Mendel (1822–1884) was regarded as such a breakthrough. Mendel was a contemporary of Darwin, but the significance of his work was not appreciated until the early twentieth century. Here at last was a mechanistic explanation for heredity that everyone had been looking for.

Thus began the study of genetics, which became an enormously sophisticated science, especially after the identification of DNA as the molecular conveyor of information by James Watson, Francis Crick, and others in the 1950s. Along the way, however, a faulty inference crept in—that genes are the *only* mechanism of inheritance. Around the world, for experts and novices alike, when you say the word "evolution" most people hear the word "genes."

Yet, if you were to ask the simple question "Are genes the *only* way that offspring resemble their parents?" even a novice would be able to answer "No." Offspring speak the same language as their parents, for example, and this has nothing to do with genes (other than the role genes play in language acquisition). Myriad other traits are also transmitted culturally rather than genetically.

Shouldn't cultural inheritance mechanisms be included along with genetic inheritance mechanisms in the study of evolution?

Clearly, to fully appreciate the relevance of an evolutionary worldview, we must think about evolution in a way that includes but also goes beyond the genetic. In this chapter I will tell three stories about other evolutionary processes: our immune systems, our capacity to learn as individuals, and our capacity to change as cultures. Each of these processes occurs far more rapidly than genetic evolution. But they are also *products* of genetic evolution. In other words, genetic evolution spawned other evolutionary processes and then coevolved with them, a process that is still ongoing. This expanded view of evolution is highly relevant to policy formulation. Not only must we see the fast-paced changes swirling all around us and even within us as evolutionary processes, but we must construct new evolutionary processes to adapt to our modern environments.

OUR IMMUNE SYSTEMS

When we die, our bodies start decomposing immediately as the microbes that are already within us run riot. The fact that this does not happen while we are living is due to our immune systems, a mind-boggling array of adaptations that evolved over hundreds of millions of years to keep our healthy microbiomes within bounds and weed out infectious agents.[1] The vertebrate immune system includes some components that are called innate because they are inherited from our parents and do not change over the course of our lifetimes. For example, when you get a splinter, your immune system cells already present at the site release chemicals that increase blood flow to the area, make the vessels more porous so that fluid from the capillaries can leak out into the tissues, and stimulate the nerves, which we subjectively feel as pain. Other chemicals, called "cytokines," diffuse outward and recruit additional cells to the site, much like the pheromones emitted by an angry wasp colony. The response will be much the same if you are eight or eighty.

Other components of the immune system are called adaptive because they are capable of rapidly changing within our bodies.

This is a good thing because the innate component alone could not keep pace with rapid microbial evolution. Here is a glimpse of how the adaptive component of the immune system works.

At this moment, about 3 billion B-cells are coursing through your veins. They are the cells that create antibodies. An antibody is a molecule that is capable of latching onto an organic surface. Any given antibody can only latch onto a narrow range of surfaces, but the B-cells produce roughly 100 million different antibodies that collectively can latch onto almost any conceivable organic surface. This is the "variation" part of an evolutionary process.

When an antibody attaches itself to a foreign object such as a bacteria, it tags the object for destruction and removal by the innate components of the immune system. At the same time, it stimulates the B-cells that created that particular antibody to multiply and ramp up production. This is the "selection" and "heredity" part of an evolutionary process. A single B-cell can divide repeatedly to become 20,000 cells within a week and each cell can pump out 2,000 antibody molecules every second. In this fashion, the antibodies capable of fighting a given infectious agent become more prevalent while the other antibodies remain at a baseline level.

So the adaptive component of the immune system is a rapid

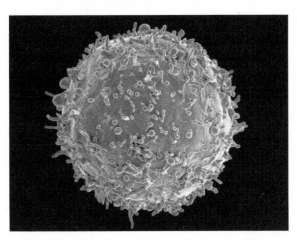

B-cells are part of the adaptive component of the immune system.

evolutionary process that includes the three ingredients of variation, selection, and heredity. Why is it important to keep this in
mind? Because it means that Tinbergen's four questions, which
organize the study of genetic evolution, can also organize the study
of the immune system. For example, when two people get the same
disease, they do not necessarily produce the same antibodies. More
than one antibody is capable of latching onto the disease and which
one gets created and amplified in any particular person is a matter
of chance, just like the adaptations that evolve in Richard Lenski's
E. coli experiment (the history question).

There are four main points to note about the immune system
for the sake of this chapter, and these same points apply in novel
ways to individual learning and cultural change. First, the ability
to create 100 million antibodies and to select the ones that bind
to antigens did not happen by a happy accident. They are elaborate products of genetic evolution operating over hundreds of millions of years. The adaptive component of the immune system is an
example of an evolutionary process built by another evolutionary
process.

Second, the adaptive component of the immune system supple-
ments and works in close association with the innate component.
For the most part, our microbiomes are kept within bounds and
infectious agents are weeded out by mechanisms that we inherited from our parents and that do not change during our lifetimes.
Antibodies play the essential but relatively modest role of tagging
the infectious agents, whose removal is left to innate mechanisms.
To understand the immune system, we need to appreciate *both* its
innate *and* its adaptive components.

Third, the immune system requires elaborate cooperation
among its component parts. Dozens of cell types must interact with
each other in just the right way. This symphony of cooperation
came about by between-organism selection. Organisms with poorly
coordinated immune systems were not among our ancestors.

Fourth, the immune system evolved in the main to increase our
survival and reproduction as individuals—but that admits exceptions. Just as the smoke detectors in our homes and intruder alarms
in our cars go off inappropriately, so do our immune systems, form-

ing antibodies against harmless substances, even when they are working according to their design. Real trouble starts when our immune systems encounter novel environments that cause them to go haywire, as we saw in chapter 3.

I have featured the adaptive component of the immune system in part because it is a fascinating and important example of rapid evolution in its own right, but also because it provides a frame of comparison for our next example.

OUR CAPACITY TO LEARN AS INDIVIDUALS

The immune system's job of fighting diseases is only one of many challenges that organisms must overcome to survive and reproduce in their environments. They must also avoid predators, find food and mates, battle the elements, and fight or cooperate with members of their own kind. Every individual's encounter with these challenges will be unique, calling for an impressive degree of behavioral flexibility. In one example, researched by Dr. Michael F. Benard at Case Western Reserve University, the tadpole stage of the Pacific Chorus Frog can inhabit one of three different environments: 1) bodies of water without predators; 2) bodies of water with fish predators, which actively pursue their prey; and 3) bodies of water with insect predators, which ambush their prey with deadly strikes of their jaws.[2] Bodies of water seldom have both types of predators because the fish eat the insects along with the tadpoles.

Each of these environments calls for different adaptations to survive and grow to the adult stage of the life cycle. The tadpoles have evolved to detect which environment they inhabit from chemical cues in the water and to express the appropriate adaptations, much as the immune system detects and responds to the presence of an invading infection. This involves not only a change in behavior, but a whole-body makeover. In the absence of predators, they move around freely in search of food. In the presence of fish predators, they remain still as much as possible and are morphologically adapted to escape in a burst of speed when detected. In the presence of insect predators, they also remain still as much as possible and

are morphologically adapted to direct the predator's attack to the
fringe of the tail, where the tadpole can get away, rather than to its
body and vital organs. This is not just speculation because Benard
could actually cause an evolutionary mismatch to take place in the
laboratory. When he raised tadpoles with chemical cues from fish
and then exposed them to insects (something that seldom happens
in nature), their mortality was greater than for tadpoles raised with
chemical cues from insects. When he raised tadpoles with chemical
cues from insects and then exposed them to fish, their mortality was
greater than for tadpoles raised with chemical cues from fish.

This kind of flexibility is like the innate component of the
immune system: a fixed repertoire of adaptations that evolved by
genetic evolution, waiting to be triggered by the appropriate environmental signals. Other forms of behavioral flexibility are more
open-ended, like the adaptive component of the immune system.
The organism behaves in ways that are more or less arbitrary (the
variation part of an evolutionary process); some ways are sensed
as more rewarding than others, and these behaviors are expressed
more frequently (the selection and inheritance parts of an evolutionary process). Thanks to this kind of open-ended behavioral
flexibility, organisms can adapt to their environments during the
course of their lifetimes in ways that can go beyond the fixed repertoire of behaviors that evolved by genetic evolution.

In many species of frogs and toads,
the tadpole stage has evolved to have a
whole-body makeover in response to
the presence and type of predator in
their environment.

The branch of psychology devoted to open-ended learning is called behaviorism, and its most famous proponent was B. F. Skinner (1904–1990), inventor of the famed Skinner box, an apparatus that enables the environmental inputs experienced by an animal to be controlled and its behavioral outputs to be recorded with scientific precision. Skinner claimed that animals could learn to do just about anything by trial-and-error learning. He even trained pigeons to play Ping-Pong by rewarding them with food pellets for scoring against their opponent—a behavior that certainly never existed during their entire previous history as a species!

So there is an important comparison to be made between the immune system as an elaborate set of adaptations that evolved to fight disease and our behavioral flexibility as an elaborate set of adaptations to surmount other environmental challenges. However, the study of learning did not proceed along the same path as the study of the immune system. What has become conventional for the latter has yet to be achieved by the former. A brief history of behaviorism will help to explain why a fully rounded "four-question" approach to the study of learning has taken so long to develop.[3]

At the beginning of the twentieth century, the physical mechanisms that cause us to behave as we do were as mysterious as the mechanisms of genetic inheritance. The techniques required for cognitive psychology and neurobiology to become sciences didn't exist yet. Speculations on how the mind worked were exactly that: speculations that relied on introspection and little else. Against this background, behaviorism originated as cutting-edge science because it enabled learning to be studied purely on the basis of environmental inputs and behavioral outputs and didn't require mechanistic knowledge about how the mind works. Stated in terms of Tinbergen's four questions, behaviorism could make progress on the function and history (meaning the history of reinforcement) questions, while largely ignoring the mechanism and development questions. On this basis, it became the dominant tradition in academic psychology during the first half of the twentieth century.

Eventually, the limitations of behaviorism became apparent. Organisms were not entirely blank slates in their learning abilities, and techniques were becoming available for studying how the mind works in a mechanistic sense. Skinner opposed many of these developments. He derided the study of mechanisms (calling it "mentalism") and overreached in his claims for what trial-and-error learning could explain. As a result, scientific progress took the form of a revolt rather than continuous change. It was called the "cognitive revolution."[4] During the second half of the twentieth century, cognitive psychologists thought of the mind as a computer, and the main objective was to understand its circuitry—Tinbergen's mechanism question.

Behaviorism had become passé within academic psychology—even taboo, because it treated the mind as a blank slate and didn't factor in the mechanisms at work in the brain—but it didn't die out entirely. Instead, it flourished in the applied branches of psychology, where the main objective is to change behaviors in the real world.

Then, starting in the 1980s, the cognitive revolutionaries were challenged by a new breed of upstarts who called themselves evolutionary psychologists.[5] The mind is not simply a single all-purpose computer, they said, but a collection of many specialized modules, each a product of genetic evolution that solves a particular adaptive problem faced by our ancestors. These "modules" are like the records in a jukebox, waiting to be played whenever the environment presses the right buttons. Even though the evolutionary psychologists challenged the cognitive revolutionaries about the mechanics of the mind, both camps were united in their scorn for behaviorism and other "blank slate" traditions in the social sciences.

Readers unfamiliar with the world of science might be surprised and a bit amused by all this Sturm und Drang. Scientists are not cold and rational like Mr. Spock on *Star Trek*. They are flesh-and-blood humans who attempt to build and defend their empires, much like people from other walks of life. The only thing that sets science apart is a set of norms and practices that results in the accumulation of factual knowledge, albeit with many twists and turns along the way. It shouldn't surprise us too much that one broad area of inquiry, such as learning, might lag behind another broad area such as immunology, even though both end up within the same theoretical framework.

What's important for our purposes is that *we* can see the similarities between our capacity to learn as individuals and the innate and adaptive components of the immune system. Here are some insights that result from the comparison. First and foremost, you can begin to think of yourself as a rapidly evolving system in your own right, adapting to your environment over the course of your lifetime. Who you are is determined in large part not by your genes (other than the genes that make open-ended learning possible), but by your environment and the behavioral options that you ended up adopting to solve the problems of your own existence. And just as you adapted to past environments, you can further adapt to your current environmental challenges. You have the capacity for positive, even transformational, change.

Yet this does not mean that your personal evolution is an "anything goes" affair, any more than you can consciously choose your own antibodies. The mechanisms that make open-ended learning possible are extremely complex, and most of them take place beneath conscious awareness. If we want to consciously direct our future evolution, we must understand and work through these mechanisms.[6]

Furthermore, as we have seen for genetic evolution, what's adaptive in the evolutionary sense of the word isn't necessarily good or right in the normative sense. Genetic evolution often results in adaptations that are good for me but not you, or us but not them, or good over the short term but not the long term. The behaviors that we adopt by open-ended learning have all the same limitations. If anything, behavioral adaptations are even more shortsighted than genetic evolution because the immediate costs and benefits of our behaviors are more perceptible to us than the long-term consequences. You might want to lose weight, but your mind is causing you to dip your hand into the next bag of Doritos. You might want peace on earth, but your mind is causing you to do what it takes to beat out your competitors for a promotion at the office. A lot of cleverness will be required to align our learning abilities to our long-term personal and societal goals.

We also must keep in mind that the adaptive component of our learning system, like the adaptive component of our immune system, works in conjunction with an innate component—a fixed repertoire of behavioral responses triggered by environmental stimuli. The first evolutionary psychologists weren't wrong to emphasize the modular nature of human and animal minds; they were just wrong to deny that such minds could also include an adaptive component.

To appreciate the innate component of our learning system, consider that all species experience both benign and harsh conditions during their evolutionary histories, resulting in conditional adaptations that are triggered by environmental signals, similar to the conditional anti-predator adaptations in the Pacific tree frog. Nonhuman species don't fall apart when times get hard—they behave in ways that are well adapted to hard times.[7] In many species of birds, stressful environments result in higher levels of corticosterone hormones in the eggs laid by females.[8] When hormone levels are experimentally manipulated in the laboratory, chicks that are equivalent in every other respect grow up to be different creatures. Those who experienced higher levels of hormones during development leave the nest at a smaller size, mature their flight muscles more quickly, and have better in-flight performance than fledglings who experienced lower hormone levels. While foraging, they are more active and are willing to take greater risks to find food. These are not deficits caused by stress, and they didn't arise by trial-and-error learning; instead, they are adaptations to stress-inducing environments that are a result of genetic evolution and lie latent in every bird, waiting to be expressed by the appropriate signal. In fact, even the open-ended learning process is tweaked. Stress-adapted birds are more likely to discount behaviors learned from their parents in favor of behaviors learned from conspecifics or from personal experience. If your mother is stressed, maybe it's because she doesn't know things that other birds know or that you can learn for yourself!

Extensive research on laboratory rats shows that stressed mothers spend less time licking their babies. When the amount of licking is manipulated in the laboratory, rat pups that are equivalent in every other respect grow up to be different creatures. Females who are licked less achieve puberty earlier, are more successful at achieving social dominance over other females, are more attractive to males, and have greater success at getting pregnant. Males who are licked less engage in more play fighting as juveniles and are more pugnacious as adults. In short, both sexes become adapted to reproduce as soon as possible, which makes good adaptive sense because in stressful environments there might be no tomorrow. This example is similar to the bird example except that the environmental signal is a maternal behavior rather than a hormone. Stressed mothers are not being neglectful by licking their pups less. If they were to lick more, then their pups would not develop the adaptations for the hard times that they are likely to encounter. In fact, female rats that are licked less as pups also lick less as mothers, even when all other aspects of their environment are held equal. From the pup's perspective, the signal transmits the experience of their grandmother in addition to their mother.

There is every reason to expect human child development to be influenced by hormonal and behavioral signals in the same way as birds and nonhuman mammals. After all, we are mammals and whatever is unique about our evolutionary history is layered on top of our more ancient pedigree. Children who experience harsh conditions such as poverty, neglectful caretakers, violence, and food shortage grow up (on average) to be different creatures than children who experience more nurturing conditions. They develop roughly the same suite of sociosexual strategies as stress-adapted rats, oriented toward early reproduction. Their open-ended learning abilities are altered. They have an enhanced ability to solve problems that result in immediate rewards as opposed to long-term rewards. At age three, they have a better memory for bad events than good events when recalling what happened during a puppet play. They are better at quickly shifting their attention from one activity to another. They perform better in high-risk situations.

These and other differences have been extensively documented in the child development literature, but their interpretation has been handicapped by lack of a modern evolutionary perspective. Instead, most developmental psychologists assume that development takes place optimally in nurturing environments and becomes impaired in harsh environments, like an automobile that breaks down under severe conditions. The challenge is therefore to fix what is broken, to make so-called at-risk children more like "normal" children. This provides an excellent example of the theory deciding what can be observed. Against the background of the "broken car" model, the idea of *behaving adaptively in stressful environments* becomes difficult to see.

Later I will elaborate on how an evolutionary worldview can lead to new practical strategies for managing our own personal evolution. For the purposes of this chapter, our capacity to learn as individuals is one of three examples intended to expand our understanding about evolution beyond the genetic.

OUR CAPACITY FOR CULTURAL CHANGE

A point I made about our immune system, which also applies to our learning system, is that it requires elaborate coordination among its component parts. This coordination is a product of between-organism selection.

In principle, information acquired by learning during one generation could be transmitted to the next generation to be refined and extended. However, this would require a high degree of coordination among individuals living in groups. Between-group selection would likely be required to encourage this degree of cooperation.

As we learned in chapter 4, between-group selection operates to some extent in nonhuman social species, but it is often strongly opposed by disruptive selection among individuals within groups, which limits group-level coordination in all its forms. For this reason, adaptations learned by individuals during the course of their lifetimes largely die with those individuals and must be learned anew by the next generation. Cultural traditions do exist in other species, but humans are clearly in a class by themselves. Our ancestors found ways to suppress disruptive competition among individuals within groups, so that between-group selection became the primary evolutionary force. This favored group-level coordination in all its forms, including the transmission of learned information across generations. Cultural evolution began to operate alongside genetic evolution and the two processes began to interact with each other.[9]

Evidence for cultural evolution is all around us, once we become attuned to it. Joseph Henrich, professor of human evolutionary biology at Harvard University, has a clever way of demonstrating its importance in his book *The Secret of Our Success: How Culture Is Driving Human Evolution, Domesticating Our Species, and Making Us Smarter.* He devotes an entire chapter to explorers who found themselves stranded in inhospitable climates, such as the Arctic, the Australian outback, or the deserts of the American Southwest. With their supplies running out, the explorers were forced to live off the land, but their individual intelligences weren't even remotely up to the task of figuring out what to eat, how to procure and prepare it, or how to protect themselves from the elements. Some of them perished and others survived only thanks to the kindness of the native people in the area who called the same inhospitable climate home. The natives were thriving, due not to their individual intelligences, but to a vast storehouse of information that had been learned by their ancestors and transmitted to the current generation without the help of a written language.

Even natives can lose their storehouse of information under some circumstances. In one famous example recounted by Henrich, the island of Tasmania off the coast of Australia used to be part of the mainland but became separated by rising sea levels approximately 12,000 years ago. The human population on the island was sufficiently small that their collective capacity to store and transmit information was diminished. Over time they led a more rudimentary existence, not because their environment was different from the mainland but because not enough heads were available to store culturally acquired information. In another example, during the 1820s an epidemic killed many of the oldest and most knowledgeable members of an Inuit population that lived in an isolated region of northwestern Greenland. The loss was like a collective stroke for the culture. The survivors were unable to make effective bows and arrows, heat-trapping entrances to their snow houses, or kayaks. They were unable to re-create this knowledge, and their population had dwindled by the time they were contacted in the 1860s by another Inuit population from around Baffin Island. Only then did the northwestern Greenland population begin to rebound, thanks to the replenished cultural toolkit obtained from another population.

Once we become attuned to it, the entire pageant of human history, starting approximately 100,000 years ago, can be seen as evolution at high speed, made possible by the transmission of learned information across generations. Our departure from Africa and colonization of the rest of the planet; our ability to inhabit all climatic zones and dozens of ecological niches as hunter-gatherers; our ability to grow food as farmers; the advent of writing; and the exploitation of fossil fuels were all made possible by the generational transfer of information.[10] The scientist and priest Pierre Teilhard de Chardin, with whom I began this book, was far ahead of his time when he asked us to imagine humanity as a twig on the tree of life that begins to proliferate so fast that it soon overtops the rest of the tree and eventually results in the coalescence of small-scale societies ("tiny grains of thought") into larger societies.

So transgenerational human cultural change counts as an evolutionary process, similar to genetic evolution, the immune system, and our capacity to learn as individuals. However, the history of thinking about culture in anthropology and sociology is at least as complex as the history of thinking about learning in psychology. Here is a brief summary to serve as a companion to my summary of behaviorism.[11]

Darwin's theory did not exist in a vacuum. It originated against the background of other prominent thinkers of the day such as Herbert Spencer (whom we met in chapter 1), Edward Burnett Tylor, and Lewis Henry Morgan. All of them believed in the unity of humankind—that people around the world are members of the same species with the same basic capacities. They also shared a progressive view of evolution, as we saw with Spencer, in which humanity is on a path toward perfection. Naturally, they thought Europeans were at the forefront, so their stance toward other cultures was to help them become more "civilized."

Darwin mostly shared these views and in any case shared the

stage with these and other major figures as the fields of anthropology and sociology emerged during the late nineteenth and early twentieth centuries. Two dissenters from these views were Franz Boas and Bronislaw Malinowski. Boas was a physicist by training who traveled to Baffin Island as a young man to study the different effects of light in the Arctic. He was so impressed by the ability of the Inuit to survive in such a harsh climate that he couldn't regard them as lower on the chain of anything. They were the best at surviving and reproducing in their particular environment, and perhaps all cultures should be regarded in the same way. Notice that this view is highly consistent with Darwin's theory of evolution; Boas was discriminating enough to distinguish between Darwinism (which he championed) and progressive forms of evolution (which he rejected).

Malinowski became intimately familiar with a native culture when he spent the duration of World War I on the Trobriand Islands in the Pacific. Like Boas, he began to appreciate the importance of viewing the world from the point of view of natives in the context of their environments, rather than placing them on a linear sequence from savagery to civilization. This led to a tradition of anthropologists living with the people whom they studied and trying to understand each culture on its own terms. The tradition self-consciously avoided any particular theoretical perspective as premature. The most important thing was to gather information as objectively as possible, which could be consulted by theories in the future. As E. E. Evans-Pritchard, a major British anthropologist during the middle of the twentieth century, described it, the whole business of anthropology is translation, to enable us to see other cultures as the members of that culture do.

This tradition in anthropology was an improvement over the arbitrary view that humanity was on a path toward perfection and led to the accumulation of a stockpile of information about cultures around the world. In the absence of a theoretical framework, however, there was no way to organize the information. The same could be said for the study of human history, which developed as a separate discipline but also became largely non-theoretical or even anti-theoretical, as if there could be no such thing as a unifying the-

Franz Boas, widely
regarded as the father
of anthropology in
America, appreciated
the adaptedness of
each culture to its
environment. Here he
is emulating the Inuit
on Baffin Island.

oretical framework. Until Darwin's revolutionary theory, the study
of natural history—the habits of plants and animals—suffered from
the same lack of organizing principles.

Traditions in cultural anthropology and other academic disciplines concerned with culture that are driven by terms such as "relativism," "social constructivism," and "postmodernism" take the rejection of theory to extremes, even to the point of denying the existence of objective knowledge altogether. The very word "theory" becomes defined as "any perspective." Science is portrayed as just another social construction, with no more or less authority than any other. Not all anthropologists go to such extremes, but the schism is so great that anthropology departments at some of the most prestigious universities, including Harvard, Stanford, and the University of California at Berkeley, have split into separate departments, like nations that split apart because their citizens just can't get along. In anthropology departments that remain intact, there is often little communication and much antipathy among the factions, as if they would like to divorce if only they could.

Fortunately, the study of human cultural diversity and change from a modern evolutionary perspective began to take root in the final decades of the twentieth century. Today there is a rapidly growing community of scientists and scholars drawn from a melting pot of academic disciplines who appreciate the meaning of having an evolutionary worldview in a social and cultural context.[12]

With this perspective, you can begin to think of yourself as not just a product of your genes, and not just a product of your personal experience, but also as one of many members of your culture who collectively contain a vast repository of information learned and passed down from previous generations. This makes you part of something larger than yourself. The information has not just been passed down, but it has also been *winnowed* through the generations, leaving us with a set of beliefs and practices that helped us to cohere as groups (Tinbergen's function and history questions).

The cultures we have inherited can be described in functional terms as *meaning systems*, which receive environmental information as input and process it in ways that result in action as output. If we exist within a well-adapted meaning system, then we arise each morning brimming with purpose and what we feel driven to do is in fact what is required to prosper. Notice that a meaning system can fail in at least two different ways. It can fail to inspire us, or it can inspire us to do the wrong things.

Meaning systems are human constructions. Hence, the tradition of social constructivism in anthropology and sociology isn't wrong, as long as it's not conceptualized as outside the orbit of science and evolutionary theory.[13] We need evolutionary theory twice over: first, to understand the genetically evolved mechanisms that make cultural evolution possible; and second, to understand the diversity of forms that results from cultural evolution.

An experiment performed on preschool children provides a glimpse of how they are able to soak up the most relevant information from their surroundings while filtering out the noise (Tinbergen's mechanism and development questions).[14] The children watched a video of two adults sitting side by side. The adults are eating different foods, drinking different-colored liquids, and manipulating a toy in different ways. Then two other adults appear and face one of the original pair. Two videos were created in which each member of the original pair was the object of attention. After watching one of the videos, the children were given a choice of which food to eat, which liquid to drink, and were given the toy to manipulate as they saw fit. The result: children were *four times* more likely to eat the food and drink the beverage and *thirteen times* more likely to play with the toy in the same way as the person who received the attention of the onlookers.

Another study of guests interviewed by Larry King, the legendary talk show host, analyzed the relationship between King and his guests. Some were socially subordinate to Larry and others were socially dominant, such as former president Bill Clinton. Acoustic analysis of the interviews showed that subordinate guests copied the speech patterns of Larry and Larry copied the speech patterns of dominant guests.[15]

These and a growing number of other studies show that both as children and adults, what we learn from others is far from random. We rely upon cues, such as who is receiving attention or who is socially dominant. You could call these copying behaviors intelligent, but it is a form of intelligence that takes place largely beneath conscious awareness. The preschoolers didn't consciously think "Oh! I will copy the person who receives the most attention from the onlookers!" and the adults didn't consciously think "Oh! I will copy the speech patterns of the person who is socially dominant!" These copying behaviors are smart in the same way that the immune system or our instincts for learning on the basis of personal experience are smart.

Mind you, we *do* consciously direct our copying behaviors to a degree, just as we consciously direct our trial-and-error learning. We also imagine and work toward social constructions that are entirely new. Henry David Thoreau had this kind of visionary planning in mind when he wrote, "If you have built castles in the air, your work need not be lost; that is where they should be. Now put the foundations under them." Nevertheless, our conscious efforts at social construction are the tip of an iceberg of mechanisms that take place beneath conscious awareness—and both need to be understood as part of Tinbergen's mechanism and development questions for the study of cultural evolution.

The products of cultural evolution (Tinbergen's function and history questions) adapt human populations to their environments much faster than genetic evolution, but they are subject to all the same limitations: at times benefitting me at your expense, us at their expense, or all of us today at the expense of future generations. To overcome these limitations, we must mindfully direct the process of cultural evolution toward planetary sustainability. As we shall see, this provides a "middle way" between laissez-faire policies on the one hand and command-and-control policies on the other—a path that anyone can follow, no matter where they currently exist on the current political landscape.

6

What All Groups Need

By now I hope you can see how an evolutionary worldview encompasses the length and breadth of human experience in addition to the biological sciences. This is something that Darwin appreciated from the beginning, but only now is the rest of the world catching on. "This View of Life," as Darwin put it, is doubly exhilarating. First, it provides a new view of the big questions pondered by deep thinkers throughout the ages, such as the nature of morality. Second, it provides new insights for improving the quality of our lives in a practical sense, from our well-being as individuals, to the myriad groups that we join to get things done, to our governments, economies, and ultimately the entire planet.

I started to explore the practical insights over a decade ago by getting involved in my hometown of Binghamton, New York, and helping to start the Evolution Institute, the first think tank to formulate public policy from a modern evolutionary perspective.[1] In the following chapters, I will recount my own experience in addition to reporting on the work of others.

You might think that the best way to report on individuals, groups, and large-scale society would be in that order. Starting at the smallest scale would make sense in the grand tradition of reductionism, which seeks to understand things by taking them apart. A version of reductionism common in the social sciences is called methodological individualism, a commitment to the belief that all social phenomena can and should be reduced to the motives and actions of individuals.[2] A version of methodological individualism common in the economics profession is called *Homo economicus*, a portrayal of individuals as motivated entirely by self-interest, usually conceptualized as the pursuit of wealth.[3]

An evolutionary worldview provides a refreshing alternative to these reductionistic traditions. Multilevel selection theory tells us that analysis should be centered on the *unit of selection*. Imagine that you are a biologist studying a solitary insect such as a fruit fly. You would study individual flies in relation to their environment to address Tinbergen's function and history questions. Then you would shift to lower levels such as organs, cells, and molecules to ask Tinbergen's mechanism and development questions.

Now imagine that you are a biologist studying a social insect such as honeybees. Since the colony is the primary unit of selection, that is the unit that you would focus on to address Tinbergen's function and history questions. You wouldn't begin at the level of individual bees any more than the fruit fly biologist would begin at the level of the fly's organs. This is a powerful refutation of methodological individualism, which states that individuals should *always* be the center of analysis.[4]

If it is true that we are a highly group-selected species, at the scale of small groups during our genetic evolution and progressively larger societies during our cultural evolution, then these groupings should be the center of our analysis, as surely as honeybee colonies are for the biologist. A sports analogy might help to make this intuitive. Suppose that you're watching a football game. The ball snaps and one of the wide receivers takes off down the field, darting first to the left and then to the right before stopping. The quarterback has thrown the ball to one of the tight ends. The wide receiver played an important role by drawing attention away from the tight end, and could have received the ball if the tight end had been covered too closely, but his behavior is impossible to understand in isolation. It only makes sense in the context of a coordinated team effort.

This example is easy to understand because we know that a sports team is strongly selected to function well as a unit. Parenthetically, when a team *doesn't* function well, it is often because of disruptive self-serving competition among members within the team. Coaches preach that there is no "I" in "TEAM" to suppress the temptation to strive for individual glory, which is felt more strongly by some team members than others. Research shows that too much disparity in the salaries of members of professional sports teams undermines their cohesion as a group.[5] Conflicts between levels of selection are on full display in the sports world, but between-team selection is strong enough to result in the team-level coordination that we see.

Multilevel selection theory tells us that something similar to team-level selection took place in our species for thousands of generations, resulting in adaptations for teamwork that are baked into the genetic architecture of our minds.[6] Absorbing this fact leads to the conclusion that *small groups are a fundamental unit of human social organization*. Individuals cannot be understood except in the context of small groups, and large-scale societies need to be seen as a kind of multicellular organism comprising small groups.

For this reason, the next three chapters are sequenced in the order groups, individuals, and large-scale societies. Groups need to come first in our analysis because well-functioning groups are required for both individual well-being and efficacious action at larger scales. If you're not surrounded by nurturing others who know you by your actions, then it will be difficult for you to thrive as an individual. If you're not part of a group that is committed to advancing a worthy cause, then you are unlikely to have the resolve and resources to advance the cause on your own.

Here are some stories that show what all groups need to function well, which is often obscured from other perspectives but makes great sense when we are equipped with the right theory.

LIN'S LEGACY

A life-changing experience for me, after deciding to apply evolutionary theory to the solution of real-world problems, was working with Elinor Ostrom, who received the Nobel Prize in economics in 2009. Lin, as she insisted everyone should call her, was a political scientist by training and largely unknown to economists at the time she won their most coveted honor. Steve Levitt, the University of Chicago economist and co-author of the bestseller *Freakonomics*, confessed that he had to look her up on *Wikipedia* and predicted that his colleagues would hate the prize going to her because it signified that "their" prize was becoming one for all of the social sciences.[7] Economics was falling off its pedestal.

What did Lin do that was so important? She studied a problem called "the tragedy of the commons," made famous by the ecologist Garrett Hardin in an article published in the journal *Science* in 1968.[8] Hardin asked the reader to imagine a village with a common pasture that was available for all of the villagers to graze their cows. The pasture can support only so many cows, but each villager has an incentive to add more of *his* cows to the herd, resulting in the tragedy of an overgrazed pasture. Hardin's example became a parable for the problem of managing common-pool resources of all sorts, such as pastures, forests, fisheries, irrigation systems, groundwater, and the atmosphere.

Economists have difficulty seeing the tragedy of the commons

A high point of my life was working with Elinor Ostrom, who received the Nobel Prize in economics in 2009.

because of their firm belief that the pursuit of individual self-interest robustly benefits the common good. When they do acknowledge the problem posed by Hardin's parable, their two main solutions are to privatize the common resource (if possible) or to impose top-down regulations.

Against this background, Lin's work was indeed revolutionary.[9] Unlike the orthodox economics establishment, which supports its ideas primarily with mathematical equations, Lin led an effort to compile and analyze a worldwide database of groups that attempt to manage common-pool resources. Some of these groups were capable of avoiding the tragedy of commons on their own, without privatization or top-down regulation. The economists were blind to something that was taking place in the real world.

An outstanding example was a group of about a hundred fishermen operating out of Turkey's coastal city of Alanya. Before the 1970s, this fishery was largely unregulated. About half of the fishermen belonged to a local producers' cooperative, but the other half could do as they pleased. Competition for the best fishing spots led to uneven use of the whole area, more uncertainty about any fisher's catch, increased production costs, and hostilities that at times escalated to violence. All of these dysfunctions can be regarded as tragedies of the commons writ large, including but also going beyond depleting the fish population.

Then a system emerged from the cooperative, was perfected over a period of years, and largely solved these problems. All licensed fishers were eligible to join the system, not just members of the cooperative. The total area being fished was divided into a number of locations spaced far enough apart so that nets set in one area would not interfere with nets set in adjacent areas. Starting every September, eligible fishers drew lots and were assigned to the named fishing locations. At periodic intervals, they rotated their locations so that each fisher had equal access to the best areas over the long term.

This arrangement was so fair that it was easy for all of the fishers to agree to it, regardless of whether they belonged to the cooperative. It saved everyone the effort of searching and fighting over sites. It was also easy to monitor, because anyone who fished where they weren't supposed to was caught out by the ones who were playing by the rules.

Not all of the groups in Lin's database managed their resources so well. Some were failing, just as this particular group of Turkish fishers were failing prior to the 1970s. Lin's great achievement was to derive eight *core design principles* (CDPs) that made the difference between success and failure. These were what all of the groups needed but only some of them had figured out for themselves.

Without further ado, here are the eight CDPs. As I list them, think about whether they might be relevant to the groups in your life.

CDP 1. STRONG GROUP IDENTITY AND UNDERSTANDING OF PURPOSE. The most successful groups knew the boundaries of their resource, who was entitled to use it, and the rights and obligations of being a group member. This was clearly the case for the Turkish fishers, who knew who was licensed and the area that they were authorized to fish.

CDP 2. PROPORTIONAL EQUIVALENCE BETWEEN BENEFITS AND COSTS. Having some members do all the work while others get the benefits is unsustainable over the long term. In the groups that functioned well, everyone did their fair share. When leaders were accorded special privileges, it was because they had special responsibilities for which they were held accountable. Unfair inequality poisons collective efforts. The system invented by the Turkish fishermen worked only because it was scrupulously fair.

CDP 3. FAIR AND INCLUSIVE DECISION-MAKING. In the groups that functioned well, everyone took part in the decision-making—if not by consensus, then by some other process recognized as fair. People hate being bossed around but will work hard to accomplish

agreed-upon goals. In addition, the best decisions often
require knowledge of local circumstances that group
members possess and top-down regulators don't. In
the case of the Turkish fishers, everyone had to agree
to the arrangement and only they were knowledgeable
enough to divide the total area into its sectors. Also,
years were required to perfect the system and only
members of the group were in a position to know what
needed adjusting.

CDP 4. MONITORING AGREED-UPON BEHAVIORS. Even when
most members of a group are well meaning, the temptation to do less and take more than one's share is always
present and a few members might try to actively game
the system. The most successful groups in Lin's database were good at detecting lapses and transgressions,
as we have seen for the Turkish fishers.

CDP 5. GRADUATED SANCTIONS. If someone isn't doing their
part, then a friendly reminder is usually sufficient to
return them to solid citizen mode—but tougher measures such as punishment and exclusion must also be
available when needed. One of Lin's favorite examples of
this and the other CDPs involved the lobster fishermen
of Maine in the United States. Like the Turkish fishers,
the lobstermen were organized into "gangs" that had
exclusive use of sections of shoreline (CDP 1). Each
lobsterman paints his buoys in a distinctive fashion so
they can monitor each other's trapping and detect the
presence of outsiders (CDP 4). When an outsider sets
traps in their area, the resident lobstermen begin the
process of graduated sanctions by tying a bow around
the buoys (CDP 5). Lin especially enjoyed telling this
part of the story. "A bow!" She would laugh. "Can you
imagine those big burly lobstermen tying bows around
the interloper's buoys!" Of course, tougher measures
will follow if the interloper doesn't take the hint and
leave the area.

These picturesque buoys allow lobstermen to monitor each other's trapping behavior and detect outsiders.

CDP 6. FAST AND FAIR CONFLICT RESOLUTION. Conflicts of interest are likely to arise in almost any group. The best groups in Lin's database had ways to resolve them quickly in a manner that was regarded as fair by all parties. That is why an organization such as a cooperative is needed, although it need not rely on outside authority.

CDP 7. LOCAL AUTONOMY. When a group is nested within a larger society, then it must be given enough authority to create its own social organization and make its own decisions, as outlined by CDPs 1–6. This was clearly the case for the Turkish fishers. When other groups in Lin's database failed to manage their common-pool resource, it was often because they weren't provided the same kind of elbow room.

CDP 8. POLYCENTRIC GOVERNANCE. In large societies that consist of many groups, relationships among groups must embody the same principles as the relationships among individuals within groups. This means that the core design principles are *scale-independent*, a point that will become important when we turn our attention to large-scale societies in chapter 8. The Turkish fishers did rely on larger-scale governance (for example, the court system) for such things as licensing and extreme conflict resolution, but in a way that contributed to the implementation of the CDPs rather than disrupting them.

Unlike orthodox economics, which is math-rich and data-poor, the CDPs are based on the study of real-world groups such as Turkish fishers and the lobster gangs of Maine. They have been affirmed by subsequent research on common-pool resource groups that followed upon Lin's pioneering studies.[10] Hence, they are highly trustworthy. Let's take a moment to reflect upon them. Notice how sensible they are. None of them are surprising, really. Another thing you might notice is how general they are likely to be. Why should they be restricted to common-pool resource groups? How about schools, neighborhoods, churches, volunteer organizations, businesses, nonprofits, and government agencies? In a sense, the very act of working together to achieve a common goal is a common-pool resource. Additional research will be required to confirm the generality of the CDPs, but intuitively it appears highly likely.

A third thing you might notice is that even though the CDPs make great sense for almost any kind of group, sadly they are not implemented by many groups. That's what Lin found for the groups in her database. Only some did a good job managing their common-pool resources. Others succumbed to the tragedy of overuse and other failures of cooperation. If the CDPs are so sensible, beneficial, and general, then why aren't they universally adopted?

Finally, even though the CDPs might seem obvious in retrospect, they weren't at all obvious to the economics profession, which is why Lin merited their highest honor. Nothing is obvious all by itself. All policies "make sense" against the background of their assumptions, but the wrong assumptions can make people blind to the importance of the CDPs. Not only was this true for economic theory and policy, but it is true for other topic areas such as education, as we will see.

I met Lin for the first time in 2009, a few months before she was awarded the Nobel Prize. That was the year of Darwin—the 200th anniversary of his birth and 150th anniversary of the publication of *Origin of Species*. Events were being held all over the world, including a workshop titled "Do Institutions Evolve?" that both Lin and I attended. It took place in a villa in the hills of Tuscany, a short distance from Florence, Italy.[11]

Although Lin was largely invisible to the economics world, her work was well known to me and my colleagues interested in human social evolution. The title of her most influential book, published in 1990, was *Governing the Commons: The Evolution of Institutions for Collective Action*. She told me that her use of the word "evolution" was mostly colloquial at the time, but that over the years she had increasingly adopted a more formal evolutionary perspective.

As we talked in the idyllic setting, I began to realize how much Lin's CDP approach dovetailed with multilevel selection theory and the saga of human genetic and cultural evolution that I have related in the last two chapters. In groups that strongly implement the CDPs, it is difficult for members to benefit themselves at the expense of each other, so that the only way to succeed is as a group. Those are the same conditions required for a major evolutionary transition, which converts a group of organisms into an organism in its own right. Lin's work didn't just follow from her school of thought within political science and her study of common-pool resource groups. It was far more general than that. It followed from the evolutionary dynamics of cooperation in all species and our own history as a highly cooperative species.

Multilevel selection theory helps to explain why the CDPs are not more widely adopted by groups. They probably would be if selection took place only at the group level, but there is always the temptation to benefit oneself at the expense of others or the group as a whole, which results in subversion of the CDPs. These efforts might be conscious or unconscious. They might even be well intentioned, as when someone feels certain that they know what's best for the group and tries to override the opinions of others (a violation of CDP 3). Externally, the core design principles can be violated by other groups (violating CDPs 7 and 8). The CDPs must be implemented strongly enough to withstand these internal and external pressures, which doesn't always happen.

Lin and I worked together for the next three years, and with her postdoctoral associate Michael Cox (now a faculty member at Dartmouth University), we wrote an academic article titled "Generalizing the Core Design Principles for the Efficacy of Groups," which placed her work on a more solid evolutionary foundation than ever before.[12]

The generalized version of the CDPs affirms the likelihood that they are needed by nearly any human group whose members are trying to work together to achieve common goals. However, this does not mean that they are sufficient. Almost all groups need the CDPs because they all need to cooperate in one way or another, but they might also need additional design principles to accomplish their particular objectives or to manage particular constraints. These can be called auxiliary design principles (ADPs), and they are as important for the groups that need them as the CDPs. For example, there are dozens of student groups at my university and all of them must be organized with a high turnover of their members in mind, since this is a fact of life for student groups. Not so for many common-pool resource groups, which can be designed with much lower turnover of their members.

Another important point that Lin stressed in her own work is the difference between a functional design principle and its implementation. Take monitoring (CDP 4) as an example. Every group needs to monitor agreed-upon behavior, but there are many different ways to monitor and some might be easier to implement than others for a given group. The fact that each principle can be implemented in different ways is similar to the interplay between Tinbergen's function and mechanism questions. In Lenski's *E. coli* experiment, each population functionally achieved the same outcome by processing glucose more efficiently, but each population achieved this by means of a different mechanism. Similarly, nearly all human groups can benefit from the CDPs (and the appropriate ADPs), but each group must find the best implementations, which can depend critically on local knowledge. This means that the design principles cannot be implemented in a cookie-cutter fashion.

Placing Lin's work on a general evolutionary foundation has profound implications for public policy. It provides a functional blueprint for any group, anywhere in the world, whose members need to work together to achieve common goals. Think of the groups in your life—your neighborhood, your school, your workplace, your church, the informal groups that you form with your friends and associates to get things done. All of them can be expected to vary in how well they accomplish their objectives, and most of them can be improved by more strongly implementing the core (and auxiliary) design principles. It is not an exaggeration to say that widespread implementation of this approach can make the world a better place—although it is important to add that relationships *among* groups (CDPs 7–8) must be managed in the same way as relationships among individuals *within* groups (CDPs 1–6). Otherwise, we are faced with the specter of groups that function well for themselves but at the expense of other groups and the larger-scale society as a whole.

Ever since my collaboration with Lin, I have been promoting the generalized core design principles approach in two ways: first, by analyzing existing information for different types of groups, similar to Lin's analysis of common-pool resource groups; and second, by working with real-world groups to improve their efficacy. Let's begin with schools, a particular kind of group tasked with educating our children.

SCHOOLS

Soon after I started working with Lin, the Binghamton City School District asked me to advise them in the creation of a "school within a school" for at-risk ninth and tenth graders that would be called the Regents Academy. I eagerly accepted their invitation as an opportunity to employ the design principles approach in my own hometown. Like many American public high schools, the Binghamton

High School attempts to provide a quality education for several thousand students, succeeding better for some than for others. My own two children fell on each side of the divide. My older daughter, Katie, graduated with top honors and went on to a top college. My younger daughter, Tamar, became so unhappy in middle school that we home-schooled her for a year and a half and then enrolled her in a Quaker boarding school in Pennsylvania, where she thrived and then went on to a top college.

Of course, many people don't have the luxury of sending their children who struggle in public school to a boarding school. And children born on the wrong side of the tracks have a very different experience than more privileged children. The "school within a school" was an attempt to help ninth and tenth graders who had failed at least three of their courses during the previous year and were very likely to drop out if nothing was done.

It was strange to review my parental experience through the lens of the core design principles. Even without making a formal study of it, I could see that the Quaker boarding school scored high. First and foremost, it was small enough for the teachers and students to know each other as individuals (CDP 1). The students were held to high standards, but they were recognized for their contributions (CDP 2). The Quaker religion is famously egalitarian and I was moved the first time I entered the 200-year-old meeting house, with the pews arranged in concentric squares around an empty cen-

Many schools are sadly lacking in the core design principles, especially for at-risk students.

tral space (CDP 3). Student performance was closely monitored, and other forms of monitoring could easily take place in a small residential community (CDP 4). Misbehaviors were flagged early and dealt with in a compassionate fashion, but persistent deviance was more severely punished and cases of expulsion were known (CDP 5). There were well-oiled conflict resolution procedures borrowed from the Quaker tradition (CDP 6). As its own independent entity, the school had maximum authority to govern its own affairs (CDP 7), and most of its relationships with other organizations appeared to be collaborative and benign (CDP 8). No wonder that Tamar blossomed there!

In contrast, the Binghamton High School appeared lacking in many of the same principles. It was too big to function as a single group where everyone knows each other. Some students had strong school spirit, but many others didn't identify with the school at all (CDP 1). There was little sense that costs and benefits were fairly distributed (CDP 2) or that students had a say in decision-making (CDP 3). Behavior was poorly monitored (CDP 4), the response to rule-breaking was inconsistent (CDP 5), and conflict resolution was anything but fast and fair (CDP 6). Authority to self-govern was compromised at all levels, from the single classroom to the entire school district (CDP 7), and relations with other groups were often dysfunctional and adversarial (CDP 8). Some activities that took place within the school, such as team sports, music, drama, and after-school clubs, did a better job at implementing the CDPs, but these were the very programs that were being cut by tight school budgets and obsessive focus on meeting state standards. I knew that in some respects Katie and her friends thrived despite the Binghamton High School and not because of it. She got a civics education when she tried to fight a top-down edict about wearing hats, for example.

Against this background, the idea that we might be able to help the kids who had flunked three or more of their classes the previous year might seem like a difficult proposition. Viewed through the lens of the core design principles, however, there were reasons for hope. The Regents Academy could accommodate about sixty students. It would have its own physical location and its own small staff consisting of a principal, four teachers, and one secretary. Although the basic school curriculum would be taught and the students would be taking the statewide Regents exams at the end of the year, the design of the school was otherwise up to us. This put us in a surprisingly good position to implement the core design principles needed for any group to function well, plus two auxiliary principles that we thought were needed in an educational context.

The first auxiliary principle was the need for a safe and secure social environment. Fear is good for escaping a dangerous situation, but it is not a state of mind conducive to long-term learning. We needed to create an atmosphere that was relaxed and playful. The second auxiliary principle was the need for long-term learning objectives to be rewarding over the short term. Nobody learns much when all the costs are in the present (e.g., boring classes) and all the benefits are in the far future (e.g., getting into college). Even gifted students fail to develop their talents if they don't enjoy what they are doing on a day-to-day basis.[13]

Guided by these ten design principles, we set about creating the best school that we could with the resources at hand. I was lucky to have one of my graduate students, Rick Kauffman, become involved with the project. He had experience as a public school teacher. I was also lucky that one of the four teachers, Carolyn Wilczynski, was a former student of mine and a good friend who was thoroughly at home with the application of evolutionary theory. Finally, the principal, Miriam Purdy, had a combination of passion, compassion, and strictness that was perfect for her role.

In addition to designing the school, we also designed a way to scientifically assess our results. Instead of recruiting 60 students for the Regents Academy, we identified 120 students who qualified (by flunking three or more classes the previous year) and randomly selected 60 of these to enter the program, while tracking the other 60 as they experienced the normal high school routine. This is called a randomized control trial and it is the gold standard for assessment. If the Regents Academy students did better than the comparison group, then we could say with confidence that our design principles approach was working. We also had access to information enabling us to compare both groups to the average Binghamton High School student.

It was easy to see why these students were struggling in school. In most cases, their lives out of school were woefully lacking in stability or nurturance. It seemed our task would be an uphill battle.

As we hoped, the students responded like hardy plants responding to water, sun, and nutrients. By the first marking quarter, they were performing much better than the comparison group. Definitive proof came at the end of the year when all of the students took the same state-mandated Regents exams. Not only did the Regents Academy students greatly outperform the comparison group, *but they performed on a par with the average Binghamton High School student*. By this metric, the deficits of years had been erased. All sex and ethnic categories posted equal gains. Everyone benefitted from the social environment that we had provided. This was only one group, but for me it was a powerful affirmation of the design principles approach.

In addition to improving academic performance, the Regents Academy rubbed off on other aspects of the students' lives. Our students had a greater sense of individual well-being and reported a greater amount of family support than the comparison group, in part because we made a point of praising the children to their parents. Most parents receive a phone call only when their kids are in trouble, but we called to say how well they were doing. The randomized control trial assured that these differences were the result of the Regents Academy, as opposed to differences that existed prior to the Regents Academy. Our students also reported liking school not only much better than the comparison group, but also *better than the average Binghamton High School student*. This is an important indication that *all* students can benefit from the core and auxiliary design principles, not just at-risk students. After all, we had merely attempted to achieve with our meager resources what Tamar's Quaker boarding school had achieved more comprehensively.

More about the Regents Academy is provided in the endnotes to this chapter.[14] Now let's expand the view to include the entire world of childhood education. A major theme of this book is "the

Like hardy plants, Regents Academy students
responded to a nurturing social environment
designed with the core design principles in mind.

theory decides what we can observe." Most educational practices
have rationales that make sense against the background of their
assumptions. It is efficient to group children into classes by age. If
our kids aren't learning well enough, it makes sense to extend the
school day and year and to eliminate "frills" such as recess and art.
Teachers should be held strictly accountable for the learning outcomes of their students. All schools should conform to a common
set of standards. No-touch rules can help to prevent physical and
sexual abuse. All of these educational policies "make sense," but
somehow they add up to a public educational system that massively
violates the core design principles.

Evolutionary theory doesn't just suggest new educational
experiments, such as the Regents Academy. It can also be used to
evaluate the vast literature on past and current educational practices, in the same way that Lin Ostrom analyzed the literature on
common-pool resource groups. Consider the Good Behavior Game
(GBG), a classroom management practice that was invented in the
1960s and has been extensively studied over a period of decades.[15]
The children in a class are asked to nominate appropriate and inappropriate behaviors, which are then displayed prominently on the

classroom walls. The lists are typically much the same as what the teacher might write—even kids who act up know what it means to be good—but it makes a big difference for the students to regard the list as *theirs*, rather than *imposed upon them*. In other words, CDP 3 (decision-making by consensus or a process otherwise recognized as fair) has been satisfied rather than violated.

Next, the class is divided into a number of groups that compete to be good while doing their class work. The competition is "soft," which means that any group can be a winner by remaining below a certain number of bad behaviors. The reward for winning can be trivial, such as picking from a prize bowl, singing a song, or even being allowed to act up for a few minutes. At first the game is played for brief periods that are announced beforehand. Gradually the duration of the game is extended and it is played unannounced until it becomes the culture of the classroom. The soft competition among groups establishes a number of the other design principles, such as a group identity (CDP 1), monitoring (CDP 4), and graduated sanctions (CDP 5). Part of the genius of the game is that the children monitor and sanction each other to make sure that their group wins the game, in addition to being monitored and sanctioned by the teacher.

Our evolutionary prediction is that all groups need the CDPs because they all need to solve the basic problems of cooperation. That goes for a class of first graders, no less than for a school for at-risk teenagers, Turkish fishers, or the lobstermen of Maine. It doesn't matter how a group comes by the CDPs. They need not be consciously designed, and most groups certainly don't have evolution in mind. They merely need to be implemented and a well-functioning group will result.

As it happens, the GBG was informed by evolution to a degree because its inventors came from the tradition of behaviorism described in the previous chapter. For the same reason, the efficacy of the game has been scientifically assessed better than almost any other classroom management technique. The figure on page 130 is from one of the first trials published in 1969.[16] Adult observers in the classroom monitored two bad behaviors that could be easily measured: "out of seat" and "blurting." During the baseline

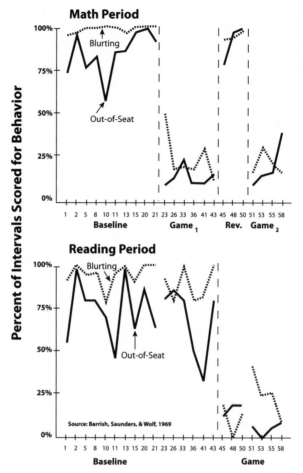

The Good Behavior Game "turns on" good
behavior almost as easily as turning on a light.

period (time interval 1–21), the class was essentially in chaos with
these behaviors being expressed between 75 percent and 100 percent of the time. When the GBG was played during the math
period but not the reading period (time interval 22–44), the bad
behaviors plummeted for the former but not the latter. Then the
GBG was played during the reading period but not the math period
(time interval 44–50), and the expression of bad behaviors flipped.
Finally, when the GBG was played during both the math and read-

ing periods (time interval 51–58), the students were well behaved in both cases. This is a good example of how scientific methods can be employed in real-world situations. Is any more proof needed that the GBG "turned on" good behavior in this particular classroom?

Unsurprisingly, a lot more learning takes place in classrooms where this technique has been implemented. More surprising is the lifetime benefits for the students. For this result we have a randomized control trial to thank that dwarfs my effort with the Regents Academy. In the 1980s, Shep Kellam, a psychiatrist at the Johns Hopkins University, persuaded the Baltimore City School District to randomly assign first- and second-grade teachers to classrooms and then to randomly assign the classrooms to one of three conditions: the GBG, another evidence-based teaching strategy called mastery learning, or no special intervention.[17] A total of 1,240 African-American students participated in the study and those who were assigned to the experimental group spent only one year in a GBG classroom (i.e., either the first or second grade but not both). The scope of this randomized control trial is impressive. Even more impressive, Kellam and his team of researchers continued to track the students into their adulthood. In fact, the study is still in progress with the participants in their midthirties.

In the two control conditions (mastery learning and no special intervention), some teachers were able to keep their class under control, but chaos reigned in the other classrooms. What good is a pedagogical method such as mastery learning when no learning of any kind is taking place? In the classes that played the GBG, the students were on task and cooperative, as expected from previous research in different locations.

You might think that the beneficial effects of the GBG would evaporate when the students went on to other classes that did not play the game. Instead, the kids built upon the social skills that they acquired during that single year. By the sixth grade, the GBG kids were less likely to face arrest or become smokers. As young adults, they were more likely to be in college or gainfully employed, less likely to be incarcerated or addicted to drugs, and less likely to commit suicide. An economic analysis conducted by an independent party estimated that for every dollar spent on the GBG, eighty-four dollars were saved through reduced special education, health care, and criminal justice costs.

An 84-to-1 benefit/cost ratio is mighty impressive. You might think that the use of the method would spread rapidly on the strength of such evidence. As it happens, the GBG is spreading worldwide,[18] but not nearly as fast as you might expect. After all, it has now been over fifty years since the game was invented. In the meantime, other educational practices have spread that flagrantly violate the core design principles because they "make sense" on the basis of other rationales. There is no evidence that modern education as a whole is gravitating toward the core design principles and no reason to expect it to, until it is viewed through the lens of the right theory. Now let's train the same lens on another kind of group.

NEIGHBORHOODS

A neighborhood is a group of people living in close proximity to each other. Neighborhoods are less important than they used to be, since it is so easy to travel beyond their boundaries, but they are still very important for child development and a high quality of adult life.[19] We know from common experience that neighborhoods vary greatly in quality, but how much of this variation can be explained by the presence and absence of the core design principles? I invite you to reflect upon this question for the neighborhoods in your own life.

One of Lin's former PhD students, Roy Oakerson, and his own student, Jer Clifton, made a study of a neighborhood association in Buffalo, New York, that was doing an excellent job implementing the CDPs.[20] Located in a run-down section of Buffalo's West Side, the West Side Community Collaborative (WSCC) began by creating small units called block clubs, where residents of a single block could meet regularly, enjoy each other's company, and accomplish manageable goals such as cleaning up vacant lots and putting in gardens. Residents identified more strongly with their block clubs than with the WSCC or the city as a whole (CDP 1). Being organized into small units meant that block club members benefitted from their own efforts (CDP 2). True, residents of a given block who did not join the club also benefitted to a degree, but they also became the target of sanctioning if they were part of the problem, such as being the owner of a derelict property. Consensus decision-making was spontaneous in very small groups that met regularly (CDP 3). The block clubs could monitor compliance with building codes much more closely than city code officers (CDP 4). When dealing with the owner of a derelict property, block club members began with friendly persuasion, but if that didn't work then city inspectors were called in (CDP 5). Mild disagreements could be discussed and settled during block club meetings or on the block. More serious disagreements could be discussed in Housing Court, which provided an independent, authoritative third party for conflict resolution (CDP 6). The block clubs were officially recognized by the City Hall's Board of Block Clubs and were free to manage their own affairs as long as they followed the general rules for frequency of meetings, election of leaders, and adoption of bylaws (CDP 7). The relationships between the block clubs and other groups, such as the WSCC and various branches of the Buffalo city government, were primarily helpful (CDP 8). Indeed, the block clubs required a larger scale of social organization (Housing Courts, code officers, and City Hall) to implement critical features of the core design principles such as graduated sanctions and conflict resolution.

I have toured the WSCC neighborhoods with Roy Oakerson and was inspired by them. There was not a randomized control trial in this case, but the brightly colored houses and profusion of flower gardens offered their own kind of proof. These neighborhoods had improved, not by an influx of money from outside the community, but by the residents empowering themselves with assistance from the city at strategically chosen moments.

Turning to the academic literature on neighborhoods, if you read classic books such as Jane Jacobs's *The Death and Life of Great American Cities* or Robert Putnam's *Bowling Alone*,[21] the core design principles leap out of almost every page. Outstanding research led by the Harvard sociologist Robert Sampson has identified collective efficacy—the ability of residents to monitor and informally police their neighborhoods (CDPs 4 and 5)—as one of the strongest predictors of neighborhood success. It is likely that a systematic review will provide strong support for all of the CDPs. Yet this does not mean that the CDPs are obvious to everyone or that more and more neighborhoods are using them over time. On the contrary, Jacobs and Putnam wrote their books because the trends were going the other way. Efforts to "develop" cities were tearing neighborhoods apart, and group activities of all sorts were becoming much less frequent than during our parents' and grandparents' generations. Thus, there is tremendous opportunity for both academic investigation and practical improvement of real-world neighborhoods by adopting the CDP approach.

Notice that in all of my examples so far, the core design principles are both old and new, familiar and ignored. They are adopted by some groups but not others and showcased by some academic schools of thought but not others. Major figures such as Lin Ostrom or Robert Putnam are famous within their academic circles but otherwise largely unknown, as we saw with Lin's invisibility to the economics profession. One of the main contributions of evolutionary theory is to establish the generality of the CDPs as fundamental to all cooperative endeavors and therefore to predict that they are needed by all groups whose members must cooperate to achieve shared goals. This bold prediction goes beyond any previous school of thought that already recognizes the importance of the core design principles within a smaller domain. Now let's see if our prediction holds for another kind of group: religious congregations.

RELIGIOUS GROUPS

If you had a religious upbringing, you might regard your congregation as among the most important groups in your life. Or you might have left your congregation behind as soon as you were able to act upon your own decisions. Religious congregations vary in how well they function for their members. Can their success or failure be explained on the basis of the same CDPs that work for common-pool resource groups, schools, and neighborhoods?

Religions would not persist if they didn't provide something for the average believer on a day-to-day basis.[22] That "something" includes a sense of meaning provided by a relationship with the divine and a supportive community of people united by the religion. These two things are closely connected. One's relationship with the divine informs one's relationships with other people. As a gospel song puts it, "If you don't love your neighbor, then you don't love God."

So a religion can do a great job creating a strong group identity and sense of purpose, which is the first CDP, but that's not enough. A religious group is faced with the same challenges as any other kind of group—a distribution of costs and benefits, decisions to be made, rules to be monitored and enforced, conflicts to be resolved, and relationships with other groups to be negotiated. If a religious group can't do these things well, then it will fall apart or be abandoned by its members for some other kind of group, which might be religious or secular.

This kind of competition is taking place all around us. My city of Binghamton includes approximately a hundred active religious congregations. Some are growing while others are shrinking. Many have folded altogether. Others are colonizing the area from elsewhere and even being invented de novo. Two major religious denominations, the Church of Jesus Christ of Latter-day Saints and the Seventh-day Adventists, originated in little towns in New York State during the nineteenth century and now have millions of adherents worldwide. Christianity and Islam, which each have billions of adherents, had the same kind of humble origins. In short, religions are great bushy trees that evolved, and continue to evolve, by cultural evolution.

The origin of Calvinism, which emerged in the city of Geneva during the Protestant Reformation, provides a great example of how the core design principles contributed to its survival and spread.[23] Although John Calvin (1509–1564) is sometimes portrayed as a religious dictator, the Ecclesiastical Ordinances that he drafted in 1541 were strongly egalitarian (CDP 3). The head of the church was not Calvin but a group of twelve pastors (including Calvin) who functioned as equals. When they couldn't agree, the decision-making circle was widened, not narrowed (CDP 6). The duty of attending to dying plague victims provides an example of how the pastors solved problems in a fair manner using mechanisms that were difficult to subvert (CDP 2). This life-threatening task was decided by lottery. Calvin was exempted from the lottery by a group decision because his death would have had a greater negative impact on the fate of the church than the death of another pastor. Thus, Calvin was recognized as a leader, but he did not have the authority to exempt himself from the lottery.

The city of Geneva was too large for everyone to know each other (although it was a mere town by modern standards) so it was divided into sectors. Each sector was overseen by an elder, whose role was to "have oversight over the life of everyone (CDP 4), to admonish amicably those whom they see to be erring or to be living a disordered life, and, where it is required, to enjoin fraternal corrections themselves and along with others (CDP 5)." Crucially, elders had to be accepted by the people being overseen (CDP 3), as well as by the church leaders and the city government (CDP 8). No wonder that the city functioned so much better after the implementation of these rules! Geneva became known as "the City on the Hill" and its social organization was widely admired and copied by other Protestant reformers.

The need to divide Geneva into smaller sectors illustrates one of the major themes of this chapter: *the small group as a fundamental unit of human social organization*. Living in small groups has been baked into our psyches by thousands of generations of genetic evolution, and small groups need to remain "cells" in the cultural evolution of larger-scale societies. One of the newest products of religious cultural evolution, the "cell ministry," illustrates the same principle. It was invented by a Korean evangelical pastor, Dr. David Yonggi Cho (he would say that it came from God), who was trying to grow his congregation and had come to the limits of his own time and effort.[24] He decided to create "cells," small groups that meet in people's homes in addition to attending the large church services. The result was so successful that his Yoido Full Gospel Church became the largest in the world, with over 730,000 members organized into more than 25,000 cell groups—a veritable multicellular organism!

Dr. Cho is a creationist, but his writing is full of biological metaphors and dovetails perfectly with an evolutionary worldview. The main power of a cell is its small size, which enables members to know each other on a face-to-face basis, just as our ancestors did in their small groups throughout our history as a species. This is in contrast to even a moderate-size church congregation, which is too large for such personal interactions. Dr. Cho has decided that fifteen families are an optimal size for a cell group. When a cell becomes larger than this, it is instructed to split.

Cell group members regard each other as family and provide the same kind of material benefits that friends and family members provide for each other, including help with the daily round of life and support during difficult times such as illness, marital problems, or losing one's job. Cell group members also provide important social and psychological benefits to each other. Dr. Cho identifies physical touching, recognition for the role that one plays, praise for exceptional contributions, and love as the most important ingredients that cell groups are exceptionally good at providing. They are also excellent for monitoring commitment to the group and conformance to agreed-upon behaviors (CDP 4). If someone doesn't attend a large church service, nothing is done about it. If someone doesn't attend a cell group meeting, there is an immediate effort to find out why and extend help if necessary.

Also, cell groups are excellent for recruiting new members. This is a good example of an auxiliary design principle, needed by some groups and not others. For most of our evolutionary history as a species, we lived in culturally homogeneous groups that did not recruit new members from other cultures. Judaism for most of its history is an example of such a cultural religious tradition. An evangelizing religion such as Christianity and Islam was a cultural innovation that changed the course of world history, creating strong and unified groups that grow not only by biological reproduction (they tend to do a good job at that too) but also by converting adults from other faiths. Today, in the modern landscape of religion, an auxiliary design principle such as the need to evangelize is as important as the core design principles. Cell groups can recruit new members from their immediate vicinity such as their neighborhoods, apartment buildings, and businesses, which is far more effective than revival meetings or knocking on doors. Some cell group leaders are so zealous that they ride up and down the elevators of their apartment complexes in an effort to recruit new members. Just as Calvinism was widely copied during the Protestant Reformation, cell group ministries have spread around the world, including some of the religious congregations in Binghamton. One Internet meme shows a smiley face surrounded by the hexagonal array of a honeycomb with the words "What the cell is to the body . . . the small group is to the church." If only religious believers knew how much their biological metaphors can be affirmed by current evolutionary science!

To summarize, our prediction that all groups need the CDPs can be extended to religious congregations, in addition to common-pool resource groups, schools, and neighborhoods. How about our workplaces?

BUSINESS GROUPS

A business, or firm, is a group of people with the shared endeavor of manufacturing a product or providing a service. It goes without saying that members of a business group must cooperate to accomplish their goals. In addition, between-group competition is fiercer in the business world than in most other walks of life. It therefore stands to reason that if the CDPs are required for groups to function well, they should have become standard business practice in the same way that they have become standard religious practice.

Yet this is very far from the truth. Instead, there is a widespread perception that the game of business hardball is played by a different set of rules, symbolized by the fictional character Gordon Gekko in the movie *Wall Street*. Greed is good, your value to society is measured by your wealth, and the only responsibility of a business is to maximize profits for its shareholders. Conventional morality isn't needed because the invisible hand of the market ensures that everything will come out right.

This worldview has become so pervasive that we have grown accustomed to it, as if it must be true, no matter how much we might wish otherwise. In reality, it is a very idiosyncratic view that has its roots in the nineteenth century but didn't become prominent until the mid-twentieth century. It is based on the particular form of reductionism known as *Homo economicus*, which imagines people as entirely self-regarding agents operating in a frictionless market.[25] Here is how the economist Herbert Gintis describes the influence of *Homo economicus* on the business world:

> After World War II, business schools blossomed all over the United States. All the major universities set up business schools. Before that, businessmen were just businessmen. They didn't go to college, or if they did they didn't learn anything about business. But these new business schools were very professional. When they wanted to teach economics, they simply borrowed from the economics discipline. In economics it's called *Homo economicus*. *Homo economicus* is not that popular any more but it certainly was after World War II. *Homo economicus* has no emotions. He's totally interested in maximizing his wealth and income. He really doesn't care about other people, although he does care about leisure. Leisure, income, and wealth are the only things. When they taught this to business school students it obviously followed that if you're a good businessman you should just maximize your material wealth. This is greedy. Being greedy is human, it's good to do, and the more greedy you are the more successful you'll be.[26]

It would be hard to imagine a more consequential case of the theory deciding what we can observe. If economists say that personal greed is the way to do business, then so it must be. Just as many educational practices that make sense against the background of their assumptions end up being disastrous for our kids, business practices that make sense against the background of *Homo economicus* end up being disastrous for our businesses and economies. That, at least, is the prediction that emerges from multilevel selection theory.

Fortunately, no matter how much the business community has

been blinded by the *Homo economicus* worldview, there is a large academic literature on business performance that can be analyzed in the same way that Lin Ostrom analyzed the literature on common-pool resource groups. Once we employ the right theory, we can see that the game of business hardball is no different from any other game requiring cooperation to accomplish shared goals. Here are some highlights.[27]

In his book *The Human Equation: Building Profits by Putting People First*,[28] the Stanford Business School professor Jeffrey Pfeffer amasses a mountain of evidence that businesses can be profitable by taking good care of their employees (CDP 2). One study reported by Pfeffer followed the fate of 136 companies over a five-year period, starting from the time that they initiated their public offering on the U.S. stock market. The management practices of the companies were coded using information from their offering prospectuses, which were publicly available. Statistically controlling for other factors, companies that placed a high value on human resources and sharing profits with employees had a much higher survival rate over the five-year period than companies that treated their employees as expendable.[29] This provides solid evidence that companies depend upon CDP 2 for their very survival.

Multilevel selection theory predicts that firms lacking the CDPs will become crippled by the disruptive self-serving strategies of employees and subunits within the organization. The sociologist Robert Jackall provides a detailed ethnography of a crippled firm in his book *Moral Mazes: The World of Corporate Managers*, which was published in 1988 and reissued in 2009 because of its relevance to the Great Recession.[30] As the jacket description puts it: "Robert Jackall takes the reader inside a topsy-turvy world where hard work does not necessarily lead to success, but sharp talk, self-promotion, powerful patrons, and sheer luck might." In this unprotected social environment, employees with higher ethical standards are either washed out of the system or learn to change their ways. Fortunately, this is not a statement about all corporations but only corporations lacking in the CDPs. As New York University's Stern Business School professor Jonathan Haidt has stressed, the entire system must be ethical for individuals to be ethical within the system.[31] The way to create an ethical system is to implement the CDPs.

In his 2013 book *Give and Take: A Revolutionary Approach to Success*, the University of Pennsylvania's Wharton Business School professor Adam Grant identifies three broad social strategies—giving, matching, and taking.[32] Givers freely help others, matchers give only when they expect to get, and takers try to get without giving whenever possible. The "greed is good" mentality suggests that givers can't possibly survive in the business world. Grant blends authoritative studies with entertaining biographies to show that givers do the best *and* the worst. They are spectacularly successful when they manage to combine forces with other givers, but they become chumps and doormats when surrounded by takers—exactly as expected from multilevel selection theory. A company that implements the CDPs is likely to become a mecca for giving, while taking is likely to accumulate in companies that lack the CDPs.

A 2012 British government report documented that employee-owned companies compared very favorably with conventional companies, especially during the 2008–2009 economic downturn.[33] Advantages of employee-owned companies include a higher degree of engagement and commitment, greater psychological well-being, and lower staff turnover, all of which make sense in terms of the CDPs. The report also identified barriers to employee ownership. These included (a) lack of awareness of the very concept; (b) lack of resources available to support employee ownership; and (c) legal, tax, and other regulatory barriers to employee ownership. In other words, employee-owned companies thrive even in a business environment that is stacked against them. How well would they do in a business environment that was stacked in their favor?

There are numerous movements to make corporations more socially responsible. One of these is B Lab (B stands for Benefit), which has established itself as a certifying organization, similar to the LEED certification for buildings.[34] Companies submit information to B Lab on their social responsibility practices in four areas: governance, workers, community, and environment. B Lab calculates a score based on this information. If it is high enough, then the company can advertise itself as a B Corp. Over 2,000 companies from over fifty nations have become B Corps, including household names such as Etsy and Patagonia. Conventional business wisdom suggests that every dollar spent on social responsibility is a dollar taken away from the bottom line. If this were true, then B Corps should be losers in the marketplace, much as we might wish otherwise. But it's not true. A 2014 study by Thomas F. Kelly and Xuijian Chen shows that B Corps are as profitable or more profitable than a matched sample of conventional businesses.[35]

Tom and Xuijian (known as Jerry to his American friends) happen to be on the faculty of my university's School of Management, providing an opportunity for me to team up with them and engage one of my graduate students, Mel Philips, to study B Corps in more detail. With the help of B Lab CEO Jay Coen Gilbert, we made site visits to five B Corps in the New York City area, where we met separately with the CEOs and groups of employees.

I must confess that before the site visits, even I was brainwashed into thinking that a genuine commitment to socially responsible goals must handicap a company in its struggle for existence against other companies. Only after talking with the employees and CEOs did I fully appreciate how B Corps became stronger, not weaker, by looking beyond their short-term profit statements. Employees were proud and often passionate about their work (CDP 1) as opposed to regarding it as "just a job." They felt that they were financially and socially recognized for their contributions and benefited from policies such as good health care, the ability to work at home, parental leave, and participation in communitarian activities while on the job, which enabled them to achieve a work-life balance (CDP 2). They were often included in making important company decisions (CDP 3). One reason that CEOs valued the B Corp certification process was not only to advertise their commitment to social responsibility, but also to monitor their company's performance in each of the four areas (CDP 4). Their commitment to employee welfare led to humane policies for enforcing agreed-upon behaviors (CDP 5), resolving conflicts (CDP 6), and giving employees elbow room in how they did their jobs (CDP 7). Finally, the companies were joining forces with other like-minded agents of the larger economic ecosystem—suppliers, customers, shareholders, competitors, and regulatory agencies—to implement the same principles at a larger scale (CDP 8), a point to which I will return in the chapter on multicellular society.

These policies are no more difficult to implement by a business corporation than any other policy—for example, a review of practices that can increase the four components of the B score, becoming inclusive about making major decisions, or implementing a work-at-home policy in which the effects on productivity are appropriately monitored. It is mostly a matter of seeing that they make sense, figuring out the best implementations, and monitoring the results. By the time we returned from our final site visit, I was personally convinced that the CDPs are needed by business groups as much as by any other kind of group. Blindness to this fact has resulted in tremendous harm at all scales, from exploited workers to an unstable worldwide economy and collateral damage to the environment. That's the bad news. The good news is the room for improvement when we view the business world through the lens of the right theory.

FROM GROUPS TO INDIVIDUALS AND THE WORLD

The bold prediction of multilevel selection theory is that all groups requiring cooperation to achieve shared goals need the same core design principles. Whether this was already obvious to you depends on how you were looking at it, either informally on the basis of your experience, or through the lens of a more formal worldview such as orthodox economic theory. As we have seen repeatedly, nothing is obvious all by itself. For every worldview that makes the CDPs appear sensible, there are others that make them appear wrong-headed or downright invisible. Hence, it is an accomplishment to establish the generality of the CDPs by showing that they follow from the evolutionary dynamics of cooperation in all species and our own history as a highly cooperative species, mostly at the scale of small groups.

No theory leads directly to the right answer, which is why predictions must be put to the test with information from the real world. Testing the prediction that what Lin showed for common-pool resource groups also holds for all other kinds of groups will occupy scholars and scientists for decades to come. My current assessment, based on my own research and review of the literature, is that the prediction is strongly supported.

But please don't wait for the experts. You can start thinking about the CDPs for the groups in your life right away. Use your own judgment and if you think that a given group can do a better job implementing the CDPs and appropriate ADPs, give it a try and see what happens. Of course, this must be done in collaboration with other members, since acting unilaterally would itself be a violation of the principles. If you would like some assistance implementing the CDPs and monitoring the results of your efforts, then visit www.prosocial.world, where you can learn more on your own and engage with a trained facilitator if you like. I will say more about this worldwide framework for working with groups in the final chapter. For now, it is important to show how strengthening your groups can help you to thrive as an individual and have a greater impact on the larger world.

7

From Groups to Individuals

What is an individual person? The fact that we have such a clear physical boundary and that all sorts of individual measurements can be taken on each of us leads easily to the wrong answer. Consider the chickens that we met in chapter 4. Hens were housed in cages and selected on the basis of their egg-laying ability. It's easy to count the number of eggs that emerge from the hind end of a hen and to regard it as an individual trait. Yet, when the most productive hen from each group was selected to breed the next generation of hens, egg productivity went down, not up, and after five generations the experiment had resulted in a breed of psychopaths.

The reason for this perverse result, as we saw in chapter 4, is that the most productive hen achieved her productivity by bullying the other hens. What seemed like an individual trait because you could measure it in an individual turned out to be the product of social interactions. In the parallel experiment, whole groups of hens were selected on the basis of their combined productivity. This experiment resulted in a breed of contented hens and an increase in productivity over the course of five generations because selecting at the group level favored cooperative rather than aggressive social interactions.

Thus, not only do the chickens serve as a parable for the problem of goodness, the focus of chapter 4, but also for the concept of *individuals as products of social interactions,* the focus of this chapter. Like the chickens, what can easily be measured about us as individuals, such as our statures, our personalities, and our physical and mental health, are not individual traits. Instead, they are the result of social processes that stretch back to before we were born— indeed, all the way back to when our distant ancestors were born if we take all evolutionary processes into account. Each person is an active participant in the social process, so there is plenty of scope for individual agency, but the idea that any one of us is "self-made" is a fiction.

In this chapter, I will show how a systemic view can help to solve many of the problems of modern civilization that manifest as individual dysfunctions. Along the way, I will introduce you to some people who first achieved distinction in their field of expertise and then discovered the added value of adopting an evolutionary worldview.

WHY WE HOLD HANDS

Meet Jim Coan, a clinical psychologist at the University of Virginia's highly regarded department of psychology.[1] Clinical psychologists are trained to work with people from all walks of life to improve their well-being, but many of them also conduct basic scientific research. Jim is a state-of-the-art neuroscientist in addition to being a clinical practitioner. Several years ago, if you had asked Jim if he accepted Darwin's theory of evolution, he would have said *"Of course!"* Yet, a patient in Jim's clinical practice caused him to rethink the entire nature of the human brain from an evolutionary perspective in a way that differed profoundly from his previous academic training.

The patient was a World War II veteran who had developed the symptoms of post-traumatic stress disorder (PTSD) in his eighties. The old soldier refused to recollect his wartime experiences or do anything else suggested by Jim. At one point he declared, "I want my wife with me." This was the first time that Jim had received such a request, but he had no reason to deny it, so the next session was with the patient and his wife. At first Jim treated her as a bystander and the patient was as resistant as before. Then she offered to hold her husband's hand. Like magic, he opened up and was receptive to therapy, but only when holding hands with his wife.

Jim was amazed and intrigued. The simple act of holding hands was having a powerful effect on the old man's behavior, which must have been mediated through brain activity. To learn more, Jim put on his neuroscience hat and embarked upon a set of brain-scanning experiments. The participants in the experiment didn't have PTSD, so Jim had to create a stress-inducing situation to serve as a rough equivalent.

If you've ever been in a functional magnetic resonance imaging (fMRI) machine, you know that it isn't much fun. You're inserted into a narrow tube that's extremely noisy. To make the experience even more distressing, Jim placed the hapless participants under threat of electric shock by attaching electrodes to their ankles. They experienced this stressful situation under three conditions: by themselves, holding the hand of a stranger, and holding the hand of a loved one. Now that Jim could look inside the brain, what did he find?

When participants of the experiment were on their own, their brains were in turmoil as various fight and flight circuits were activated. Holding the hand of a stranger had a slight calming effect, but the most dramatic calming effect came from holding the hand of a loved one. Jim had succeeded at duplicating in the laboratory what he had observed with the old soldier and his wife.[2]

As with many basic science experiments, this one is likely to produce a reaction of "Amazing!" followed quickly by "But didn't we already know this?" Jim's mother even scolded him for spending so much time and money when he could have just checked with her! Nevertheless, Jim was breaking new ground in a number of ways. If the results are so obvious, then why haven't clinical psychologists incorporated it into their practice? Why do they assume that the *individual* is the unit of treatment? Why was Jim surprised by the old soldier's request to have his wife present and doubly surprised when the physical act of holding hands was required? Jim's clinical and academic training had blinded him to the calming effect of holding hands, however obvious it might have been to his mother.

Perhaps most important, Jim's experiments were beginning to address what actually goes on in the brain when a person feels socially connected or isolated in a stressful situation. This was new for everyone, including Jim's mother. Yet Jim's efforts to understand the brain's activity patterns were going nowhere until one of his departmental colleagues, Dennis Proffitt, offered two pieces of advice: First, think of the person holding hands as the normal condition, not the person facing the situation alone. Second, read *An Introduction to Behavioural Ecology* by Nicholas B. Davies, John R. Krebs, and Stuart A. West.[3]

According to Jim, taking this advice revolutionized his understanding of the brain and allowed him to make sense of his own experimental results for the first time. Here is how Jim described to me in a 2016 interview the impact of reading *An Introduction to Behavioural Ecology*.

I think "meteoric" is the right way to describe its impact on me. By the time I had finished the first chapter I was already thinking about my own work, and indeed, thinking about psychology as a broad discipline completely differently. The book starts out introducing principles that organize behavior—that when you give them even a little bit of thought make complete sense. Principles like the management of bio-energetic resources; that if you're going to engage in a behavior as an organism, to accrue resources, you have to invest resources that you have in store. That is a very risky business, so you need a certain amount of information about the demand of the environment and your own resource cache. That entails certain principles that get built into the genome over time about keeping an excess, having a surplus, and maintaining a surplus.

I had a kind of personal and intellectual crisis, where I thought "Holy shit! What have I been doing all this time? I've been thinking about constructs that aren't tethered

to any ultimate goals or any ultimate constraining principles." In psychology, anything goes, because the thinking
isn't constrained by these imperatives of biological organisms across evolution and ontogeny. Then I started going
through chapter after chapter, example after example, of
these principles existing not just as logical arguments but as
empirical data. It was enough to almost make me cry.[4]

Jim's conversion experience will sound strange to many people
who assume that because an academic discipline such as psychology
or neuroscience is sophisticated in *some* respects, it must be sophisticated in *all* respects. In fact, as broad disciplines, they have done
a poor job addressing Tinbergen's function (how our psychological and neurological traits contribute to survival and reproduction)
and history (the particular evolutionary trajectory of our species)
questions, as strange as this might seem. When these two questions
are poorly addressed, then research on Tinbergen's mechanism and
development questions often heads in unprofitable directions.

An Introduction to Behavioural Ecology filled in the function and
history questions for Jim. The first edition was published in 1981
and helped to consolidate Tinbergen's four-question approach, as I
recounted at the beginning of this book. It showed that if we want
to understand how a given species behaves, we must understand
how its behaviors were shaped by natural selection. This, in turn,
requires understanding the species in relation to its ancestral environment (which might or might not be different from its current
environment) and the historical contingencies of its evolution. Asking the simple question "How would this species behave if it were
well adapted to its environment?" leads to hypotheses that provide
excellent starting points for scientific inquiry, even if not all of them
prove to be correct. This is called "adaptationist" or "natural selection" thinking, and it is one of the most powerful tools in the evolutionary toolkit, as I showed in chapter 2.

Natural selection almost always involves trade-offs, based on
the familiar principle that a jack-of-all-trades is a master of none.
Doing one thing well requires doing other things poorly. The shell
of a turtle protects it against predators but also prevents it from

moving fast. Penguins are agile swimmers but look ridiculous as they waddle about on land. Pigmented skin protects against ultra-violet radiation but prevents the synthesis of vitamin D. Every species is a bundle of trade-offs, based on the selection pressures acting upon it over evolutionary time.

The way an organism behaves during its lifetime also reflects trade-offs. A turtle's shell is an anatomical trait. A turtle pulling into its shell when it senses danger and coming out of its shell when the danger passes are behaviors. Both can be understood in terms of trade-offs. The idea that behaviors evolve, just like anatomical traits, was the point made by Niko Tinbergen and came to fruition with *An Introduction to Behavioural Ecology*. Yet such is the isolation among academic disciplines that these developments in the study of animal behavior didn't spread to the study of human psychology and neuroscience, at least as far as Jim's training was concerned. That's what he meant when he said during our interview: "In psychology, anything goes, because the thinking isn't constrained by these imperatives of biological organisms across evolution and ontogeny."

As a born-again behavioral ecologist, Jim became obsessed with the question "What is the ecology that humans are adapted for?" Here's how he described his quest in our interview.

> As I kept pulling on that thread, the only thing that kept coming up over and over again was "other humans." This is part of the model that we developed called the social baseline model. When you average everything out that humans have experienced, over millennia, the only thing that's constant is other humans. We're so adaptable, we've been to so many places; we live in the arctic and the equator. We eat whale blubber and unrefined grains. We've been to the moon and practically at the bottom of the ocean. The only thing that's constantly there is other people. I don't think I could have had that insight if I hadn't started depending upon evolution and behavioral ecology as a framework for forming questions and for asking what comes next.

It's not just other people that have been a constant throughout our evolutionary history, but other *cooperative* people in close-knit groups. Don't get me wrong—conflict has taken place throughout human history, but much of it has been conflict between groups, which means that individuals are cooperating with members of their own groups. Jim's central insight was that the human ancestral environment was to be surrounded by cooperative others and that this is reflected in the design and mechanisms of the human body and brain. That's what Jim's colleague meant when he said that holding hands was the normal condition and that facing a stressful situation alone was abnormal as far as the adaptive design of the brain and body is concerned.

In Jim's social baseline model, when the human brain and body make trade-off decisions, they automatically take social and personal resources into account. To see how this works, imagine that you are standing at the base of a long hill. How steep does it look and what is your willingness to climb it? These might seem like separate questions, but a series of ingenious experiments by Dennis Proffitt, Jim's departmental colleague, shows that they are entwined in our minds.[5] Proffitt had people estimate the slope of a hill under various conditions, such as with and without a heavy backpack, with and without a prior period of fasting, or with and without a prior physical workout. It makes sense that people would be less willing to climb the hill with a heavy backpack, after a period of fasting, or after a workout, but the surprising result was that they *estimated the slope of the hill to be steeper* under these conditions. Their willingness to climb the hill influenced their very perception of the hill.

These examples show how our brains take our personal resources into account when making trade-off decisions. But Dennis also had people estimate the slope of the hill when they were alone compared to standing next to a friend. The mere presence of a friend caused the slope of the hill to appear less steep. The brain had taken a social resource into account (the presence of a friend) as effortlessly and unconsciously as taking personal resources into account (carrying a heavy backpack, fasting, and having already had a workout).

How steep is this hill?

I have already made the point in chapter 6 that small groups are a fundamental unit of human social organization, required for individual well-being and efficacious action at a larger scale. Jim's social baseline model shows that the need to be surrounded by cooperative others is deeply written into our brains and bodies. We are hardwired to live in the presence of cooperative others, and facing the world alone places us in a state of alarm. Chronic aloneness is injurious to our mental and physical health. This is paradigmatically different from regarding individuals as the basic building block of human societies, which underpins modern economic theory and other forms of individualism.

The practical import of Jim's research is clear: If you want to improve your own well-being, then surround yourself with cooperative others. The role of handholding and touching per se is a fascinating topic for future research. Is it possible that the brain relies upon tactile contact to assess social support and that a cooperative group without touching won't work as well as a cooperative group with touching?[6] If so, that might cause us to reconsider well-meaning but misguided "no touch" rules in schools and workplaces. Only future research can answer these and other questions that are vital for individual well-being.

THE NURTURE EFFECT

Tony Biglan is a senior scientist at the Oregon Research Institute, an organization that has been working to understand human behavior and improve the quality of life since 1960. In his seventies and still in robust health, Tony has "been there and done that" with research on a panoply of problems that manifest themselves as destructive behaviors such as substance abuse, obesity, delinquency, and early sexual activity. Tony's research has been funded through numerous branches of the National Institutes of Health, including the National Cancer Institute, the National Institute on Drug Abuse, the National Institute of Mental Health, and the National Institute of Child Health and Human Development. He is a former president of the Society for Prevention Research and served on a committee convened by the Institute of Medicine, a part of the National Academy of Sciences, which reviewed the progress that the United States has made in preventing these problems. Few people have a broader understanding of the sprawling world of the behavioral health sciences than Tony.

Tony has synthesized this knowledge for a general audience in his book *The Nurture Effect: How the Science of Human Behavior Can Improve Our Lives and Our World*, which was published in 2015. The message of Tony's book is disarmingly simple: Prosociality—people nurturing other people—is a master variable. Those who are surrounded by helpful others develop multiple assets. Those who are surrounded by indifferent or hostile others develop multiple liabilities. Of course, this is where Jim Coan's research also points. If we could say only one thing about making the world a better place, to be reflected in our social policies and our personal decisions, it would be to increase nurturance throughout the life span and especially during its early stages, starting before birth.

The novelty of this message is illustrated by all of those institutes that have supported the research of Tony and his colleagues over the years. They signify that the behavioral health sciences are divided into research communities that treat problems in isolation. The majority of behavioral health scientists are not like Tony. They spend their careers studying a single problem behavior and have a hard time seeing the forest for the trees. In fact, there is little incentive to see the forest when your next grant comes from an organization that wants you to see your particular tree in ever greater detail.

Tony escaped this fate in part because he was influenced early in his career by the tradition of behaviorism and its most famous proponent, B. F. Skinner. As I recounted in chapter 5, when behaviorism was purged from academic psychology by the cognitive revolution, it didn't go extinct but continued to thrive in the applied behavioral sciences. Behaviorism's central insight for practical change efforts is captured by the phrase "selection by consequences."[7] When we act a given way, our bodies and minds are wired to sense whether the consequences are positive or negative. Did the action produce pleasure or pain? Did it cause others to smile or frown? Based on this feedback, which is automatic and takes place largely beneath conscious awareness, we either ramp up or tamp down the expression of the behavior the next time around. In this fashion, we adapt as individuals to our environments in much the same way as populations adapt to their environments by genetic evolution.

Behaviorism might have fallen out of fashion in academic psychology, but for people such as Tony trying to solve behavioral problems in the real world, it was too useful to be discarded. The fact is that our behaviors *are* shaped by their consequences to a large degree—and the environments that select our behaviors are in large part our social environments. As we have seen, we lived in small and highly cooperative groups for most of our evolutionary history. We received a lot of nurturance and were expected to give in return. If we tried to boss others around or do less than our share, we quickly received social feedback to mend our ways. The more we contributed to common goals, the more social approval and material benefits we received. This made succeeding at the expense of others a dangerous game and succeeding by working with others the surest way to survive and reproduce as an individual. We are genetically adapted to crave social acceptance and will do almost anything to achieve it.

Even so, nurturance within groups is not the whole picture of our evolutionary history. Succeeding at the expense of others might be a dangerous game, but it's still a game worth playing under some conditions. Genetic evolution has endowed us with skills for self-ishness in addition to skills for cooperation. Consider the most intimate bond that can be found in mammals, the relationship between a mother and her offspring. Even this bond is not entirely nurturing. Mothers are shaped by genetic evolution to maximize their total reproductive success over the course of their lifetimes, not the success of any particular offspring. Offspring are shaped by genetic evolution to value their own welfare over the welfare of their mother and siblings. Fathers in many mammalian species are shaped by genetic evolution to maximize their reproductive success by mating with as many females as possible and not con-tributing to childcare at all. There is even an evolutionary logic to infanticide—the killing of current offspring to have other offspring in the future.[8]

Conflict between mother and child begins before birth, as the fetus seeks to extract more resources from the mother than she is necessarily inclined to give.[9] The mother-offspring interaction at this stage of the life cycle is a biochemical tug-of-war that isn't even mental. After birth, human parents and "alloparents" (a catch-all term for anyone who participates in childcare)[10] are genetically prepared to withhold nurturance from children under harsh condi-tions, and children are genetically prepared to make the best of a bad job when they are not receiving the nurturance that they need. Many of these physiological and neural mechanisms evolved in the mammalian lineage long before we existed as a species and even before primates existed as a branch of the mammalian tree.

To make things even more complex, social interactions among groups is a different matter than social interactions within groups, as we saw in chapter 4. A group whose members nurture each other might have amicable or hostile relations with other groups, depending upon the circumstances. Amicable relations such as trade and the exchange of marriage partners extend far back in our evolutionary history,[11] but so do hostile relations such as raiding, cannibalism, and the all-out extermination of other groups.[12] Our ability to make distinctions between "Us" and "Them" and to confine our nurturance to "Us" might seem paradoxical and hypocritical from some perspectives, but it is only to be expected from an evolutionary perspective.

Once our ability to evolve as individuals was coupled with the ability to transmit learned information across generations, then human evolution flipped into warp drive, as I recounted in chapter 5. We spread over the globe, adapting as small-scale societies to all climatic zones and dozens of ecological niches. Then, with the invention of agriculture about ten thousand years ago, we expanded into the mega-societies of today. That's enough time for some but not a lot of genetic evolution. Our ability to function as members of large-scale societies, and the ability of these societies to function as well as they do, testifies to the open-ended behavioral flexibility of our species, which is the heart of the behaviorist tradition.

The story that I have just recapped is not only evolutionary, it also invokes evolution in many ways and at many different timescales. How does someone like Tony use this overarching theoretical framework to address behavioral problems in the modern world? One basic prescription is to do everything possible to re-create the ancestral social environment of small groups of nurturing individuals who know each other by their actions. Provide such an environment, and prosocial child development and adult relations will take place with surprising ease. In the absence of a nurturing social environment, the shaping of behavior will lead in a very different direction—survival and reproductive strategies that are predicated on the absence of social support; that benefit me and not you, us and not them, today without regard for tomorrow. That's what Tony means by calling nurturance a master variable.

More insight can be gained by taking seriously the concept of an individual as an evolving system. If you are getting the hang of evolutionary thinking, then you know that what counts as adaptive in the evolutionary sense of the word is not necessarily adaptive in the normative sense of the word. This is because what's normatively adaptive is typically good for everyone over the long term, whereas what's evolutionarily adaptive is often good for only a subset of everyone, often at the expense of others and the whole, and it often only leads to short-term gain without regard for long-term consequences. The situation is not hopeless—social environments can be constructed that align adaptation in the evolutionary sense with adaptation in the normative sense—but work informed by knowledge of evolution is required. Here is an example in an everyday setting that most of us can relate to.

Imagine that you're in a supermarket observing a small child having a tantrum. Maybe the mom gives in or maybe she gives her kid a verbal or physical licking. Either way, nobody is having any fun, including the onlookers. The brilliance of seeing this mundane problem from an evolutionary perspective is to realize that both the mom and the kid were behaving adaptively in the evolutionary sense of the word. Every time kids get their way by behaving obnoxiously, they are more likely to repeat it the next time around because their behavior has been reinforced. Every time the parents get their way by delivering a verbal or physical licking, their behavior is also reinforced, so they will likely repeat their behavior the next time around. The parents and their kids are locked in a behavioral coevolutionary arms race, similar to the genetic arms race between cheetahs and gazelles to run ever faster. Never mind that the behaviors aren't good for the family or society as a whole. Evolution is relentlessly relative and the family members are adopting the behaviors that maximize their relative advantage within the family. It's not even necessarily conscious on their part—what's conscious is just the tip of the iceberg of what takes place. In Skinnerian terms, it's just a form of personal evolution that happens to be poorly aligned with normative goals. Call it "the tragedy of the family commons."

You can escape a tantrum in a supermarket just by changing aisles (unless it's your kid), but you can't escape the societal conse-

Before we try to solve the problems
of world peace, let's see if we can
solve this one . . .

quences of kids whose social skills have been honed to a fine edge
by countless iterations of obnoxious interactions with their parents
and other caregivers. When these kids go to school, they don't
know how to behave differently and are likely to trigger negative
reactions in their classmates and teachers. In response to rejection,
they are likely to ramp up their obnoxiousness. It's how they've
been reinforced to respond. As they grow, they are likely to fall in
with others who deploy the same social strategies, reinforcing each
other's behaviors in deviant peer groups, similar to their own families. As adults, they might have trouble finding or keeping work,
increasing their likelihood of engaging in criminal activities. As
parents, they are likely to perpetuate the cycle.[13] Coevolutionary
arms races such as this can be highly stable. The fact that they are
socially pathological in the normative sense of the word is beside
the point. Nature is replete with similar examples, such as cancer

cells that destroy multicellular organisms and species such as the crown of thorns starfish that bring down entire coral reef ecosystems. Evolution does not make everything nice!

Yet most of this havoc can be avoided with a little evolutionary know-how. "Selection by consequences" need not be a race to the bottom. It can also result in a virtuous coevolutionary spiral. The golden rule is "Abundant rewards for good behavior, coupled with mild punishment for bad behavior that escalates only when necessary." Note that this accords with Elinor Ostrom's fifth core design principle, concerning graduated sanctions, which are needed for families along with all other kinds of groups. Also, it is important for the reinforcement to come from meaningful others, such as respected adults and well-liked peers. From these people, rewards and punishments can be as simple as a smile and a frown.

Many parents and caregivers of children employ these rules spontaneously, without needing to be taught. Others need to be coached, or more accurately, their behaviors need to be shaped by a new set of consequences in the form of an active intervention. Those reality shows where families from hell are visited by Super Nanny are not entirely divorced from reality, even though they are highly scripted. To learn more about the science behind the Super Nanny shows, I recommend visiting the website of Triple P, which stands for "Positive Parenting Program."[14] Triple P makes the scientific research conducted by people like Tony available to families around the world. Coaching tips that you can learn include "planned ignoring," "spending quality time," and "creating a learning environment." Doing nothing in response to an unwanted behavior prompts a child to try something else—perhaps a more positive behavior that can be reinforced, especially when you are spending quality time and there is something cool and interesting to be learned. "Time out" is one of the tips, but it is only employed when other efforts to reinforce good behavior have failed. When family life is running smoothly, it doesn't need to be triggered at all.

If these coaching tips seem manipulative, remember that we are always manipulating each other in one way or another. That's what it means for a person to be a product of social interactions. The reason that the word "manipulation" sounds sinister, in contrast to the more neutral phrase "social interaction," is because it implies influencing the behavior of others in a self-serving fashion and against their interests. This particular connotation of the word is indeed something to be avoided and what the CDPs discussed in chapter 6 protect against. Even small children benefit from social environments governed by the CDPs, as we saw with the Good Behavior Game, and this can be true for families as well as schools.

There is nothing sinister about the testimonials that can be found on the Triple P website. One divorced father expressed gratitude that he was now able to take his eight-year-old son to Disneyland without any behavioral meltdowns. Before he received coaching from Triple P, "I was having a lot of difficulty with him, getting him to do what I wanted him to do, and listen to me." Afterward he reported his relationship with his son as "from chaos to peace."

As a highly science-based method, Triple P has gone way beyond testimonials and working with single families. Consider a massive study conducted by Ron Prinz, Carolina Distinguished Professor of Psychology at the University of South Carolina and one of Tony's colleagues.[15] Ron selected eighteen South Carolina counties that were roughly comparable in terms of size and demographics. Nine of these counties were randomly chosen to receive coaching in Triple P. The other nine counties formed a comparison group. This was a randomized control design, similar to what Rick Kauffman and I did with the Regents Academy described in the last chapter but on a much larger scale.

How do you coach an entire county in Triple P? Ron employed a multilevel approach. Level 1 used mass media to reach all parents, no matter what their parenting style. Level 2 provided advice to parents from childcare providers and social service workers who frequently contact parents, in the form of brief consultations and ninety-minute seminars. Level 3 provided more intensive training for the parents of children exhibiting problem behaviors. Additional levels provided yet more training for families dealing with a wider range of issues, such as parental depression and marital discord. The goal in all cases was to get parents off the negative spiral of reinforcing obnoxious behaviors and onto the positive spiral of reinforcing prosocial behaviors. One advantage of this multilevel approach is that the families who need the intervention the most are not stigmatized. They are taking advice that is being offered to everyone.

Implementing all of this in nine counties while tracking outcomes in eighteen counties was a big job—but the results were worth it. Using the comparison group as a baseline, counties that implemented Triple P experienced fewer cases of substantiated child abuse, fewer out-of-home placements due to child abuse problems, and fewer hospital-reported child injuries. Stated in raw numbers, in a community of 100,000 people, Triple P resulted in 688 fewer cases of child abuse, 240 fewer out-of-home placements, and 60 fewer children needing hospitalization. Using very conservative estimates of cost-effectiveness by an independent assessor, every dollar spent on Triple P saved over six dollars that would have been spent attempting to address the problems that Triple P prevented.

Implementing Triple P wasn't easy, but it was the kind of hard work that has a reliable payoff at the end—and this is only one of the intervention programs with proven results that Tony describes in *The Nurture Effect*. According to Tony, *we already know what works*, once we view individuals as products of social interactions from an evolutionary perspective.

NOT MY MOTHER'S THERAPY

Meet Steven C. Hayes, Nevada Foundation Professor of Psychology at the University of Nevada at Reno. Steve is one of the most prolific and widely cited psychologists in the world, having authored over 35 books and 500 academic articles. He is best known for a type of psychotherapy called Acceptance and Commitment Therapy (ACT, pronounced as one word), which was featured in a six-page article in *Time* magazine in 2006. His ACT self-help book *Get Out of Your Mind and into Your Life* has sold over half a million copies.[16]

For many people—including myself before meeting Steve—psychotherapy has a reputation for being slow, inefficient, and unscientific. My mother was a devoted follower of Freudian psychoanalysis and visited her analyst once a week for most of her adult life. She felt that she gained from the experience, but when this kind of therapy was scientifically assessed, it proved to be no more effective than talking with one's pastor or a sympathetic friend.[17] Freud was a medical doctor but his methods weren't even remotely scientific. And while he was clearly on to something with his theory of the unconscious mind, there were no techniques that could take him beyond sheer speculation.

In contrast, consider one of Steve's publications.[18] Foreign students who study in America can become highly stressed, which is manifested as anxiety and depression. Steve and his colleague recruited seventy Japanese students studying at the University of Nevada and randomly assigned them to two groups. The first group read Steve's self-help book and worked through its exercises. The second group did the same after a two-month waiting period. Their physical and mental health was monitored with widely used assess-

Sigmund Freud,
an icon of the
twentieth century,
requires substantial
updating from a
modern evolutionary
perspective.

ment tools, such as the General Health Questionnaire (GHQ) and Depression Anxiety Stress Scales (DASS).

The Japanese students who volunteered for the study were highly stressed, with nearly 80 percent exceeding clinical cutoffs on one or more measures of the DASS. Individuals in both groups, on average, became less stressed after reading Steve's book and working through its exercises. The fact that the second group didn't improve until two months after the first group identifies Steve's book as the causative factor, ruling out unmeasured factors that would have influenced both groups at the same time. This was the reason for randomly assigning the participants to two groups and having them begin at different times.

Using the GHQ and DASS enabled Steve to compare his results to many other studies that used the same assessment tools. Based on this comparison, Steve can say with authority that "bibliotherapy"—reading his book without any therapist at all— provides about 30 percent of the benefit of seeing a trained ACT therapist. In a second study, he achieved similar results with public school teachers who were suffering from burnout.[19] Most self-help books boast on their covers that they will cure what ails you, but Steve can *prove* it!

If you read Steve's other scientific articles, along with hundreds of studies by other authors, you'll see that psychotherapy can be fast, efficient, and scientific. Even better, the "T" in ACT can stand for "Training" in addition to "Therapy." No matter what your current level of functioning, you can probably benefit from ACT and related techniques, just as athletes at all skill levels benefit from coaching.

What ingredients make ACT capable of achieving such impressive results? Increasingly, Steve is describing ACT as a managed process of personal evolution.[20] To see how this works, imagine someone in desperate mental circumstances who looks upon "normal" people with envy. We are accustomed to thinking of such a person as abnormal, as if their brain is not functioning as it should. Now consider the possibility that there is nothing organically wrong with this person, by which I mean that they have the same basic mental equipment as everyone else. Their only problem is that evolution has taken them where they don't want to go and that's what is causing them discomfort. With a little know-how, evolution can also become their solution.[21]

As with all forms of evolution, our personal evolution requires variation and selection. If we don't behave in different ways, then by definition we are stuck doing the same thing. If we do behave in different ways, then our actions will typically have different consequences, which we will sense in terms of pleasure or pain, social approval or rejection, and so on. Based on these consequences, the complicated machinery that genetic evolution endowed us with will cause us to ramp up the most rewarding behaviors and tamp down the most punishing behaviors the next time around. We probably won't know that this is happening, because most of the mechanisms take place beneath conscious awareness.

A visual metaphor called an adaptive landscape, which has a venerable history in evolutionary thought, can explain how someone's personal evolutionary process can trap them in desperate circumstances.[22] Imagine that you inhabit a land with many hills and valleys, with some hills much higher than others. Your only desire is to climb upward. The higher you go, the happier you become. You therefore start merrily climbing the slope of whatever hill you are on. If it is the slope of a tall hill, you become very happy indeed. If it is the slope of a short hill, you get to the top and look around you with despair. You can see taller hills in the distance, but every step that you take brings you down.

Welcome to the world of the depressed person, who finds staying in bed more rewarding than facing the day. Or the alcoholic, who finds the next drink more rewarding than abstinence. Or the person with an anxiety disorder, who will do anything to avoid a panic attack. All are behaving adaptively, by choosing the behavior that is most rewarding compared to the alternatives. The problem is that they are standing on top of very short hills.

Fortunately, there is a solution to this problem, because our perception of up and down depends on how we think about it. This is one way that ACT includes but goes beyond the behaviorist tradition of B. F. Skinner and even begins to incorporate the valid elements of Freud's thought. We are highly distinctive and perhaps even unique in our capacity for symbolic thought. In almost all other species, mental associations are closely related to environmental associations. A rat will associate cheese with the word "cheese" only as long as you pair the two with each other. In contrast, if I say the word "cheese" to you a million times without presenting you with cheese, you might smack me but you won't forget the association. We even have words for things such as "troll" that don't exist in the real world!

In short, all of us live in a symbolic world inside our heads in addition to an external world. One of Freud's great contributions was to recognize and begin to explore this inner space, whose very existence poses an evolutionary puzzle. What good is a symbolic world if it doesn't directly correspond to the external world? The answer, from a modern evolutionary perspective, is that every set of symbolic relations inside our heads results in a suite of actions that take place in the external world. Trolls might not exist, but belief in them alters behavior. More generally, if we call your particular symbolic world your "symbotype" and your measurable behaviors your "phenotype," then there is a "symbotype-phenotype relationship." This wording is useful because it provides a parallel to the phrase "genotype-phenotype relationship" used for the study of genetic evolution, where each person's set of genes (their genotype) results in a corresponding suite of measurable traits (their phenotype).[23]

This comparison, between our genes and our symbols, is full of implications for psychotherapy. Consider how we already think about genes. Species are different from each other because they possess different genes. At the same time, in sexually reproducing species every individual is genetically unique because of recombination. Despite the immense complexity of gene-gene and gene-environment interactions, a single mutation can result in a different phenotype. We are beginning to exploit this fact with so-called gene therapy, in which we surgically alter our genotypes to cure diseases or enhance our abilities. If symbotypes are like genotypes, then the same possibilities exist for psychotherapy—and we can find evidence by looking at the psychological literature in the right way.

As one example, imagine that you are a college student taking a large introductory psychology class. The professor gives you a curious assignment: Spend fifteen minutes writing about matters of importance to your life. Do this three separate times spaced a week apart. Unbeknownst to you, only half of the class received this assignment. The other half (randomly assigned, of course) was told to write on comparatively neutral topics such as sports or current events for an equivalent amount of time.

This experiment has been performed numerous times by James W. Pennebaker, a health psychologist at the University of Texas.[24] The results are astounding. Compared to the students who write about neutral topics, students who write about matters of importance to their lives get better grades and get sick less often over the course of the semester. Forty-five minutes of self-counseling, without the help of any expert, causes them to think about their lives in a way that alters their personal evolution for the better.

Here is another example. Imagine that you're a college freshman. Your new life is exciting but also filled with uncertainty: Will you make friends? Will your grades be as good as they were in high school? You take part in a study that informs you about a survey of college seniors reflecting upon their freshman experience. The seniors reported that they felt uncertain, just like you, but they soon adjusted and everything worked out fine. This was the experience of both genders and all ethnic groups, according to the survey.

Next, you are asked to write a short essay describing how your own experience echoes the results of the survey and to turn the essay into a speech that you read in front of a video camera, to be shown to future students to ease their transition to college. All of this requires one hour of your time. During every day for the next week, you are asked to complete a short survey reporting your experiences and how you felt about them, including your sense of belonging. Unbeknownst to you, students in a control group went through the same procedure with the exception that the survey was about sociopolitical attitudes rather than adjusting to college.

This experiment was performed by Gregory M. Walton and Geoffrey L. Cohen at Stanford University.[25] The essay and speech were clever ways of getting the students to internalize the results of the survey. Three years later, when the participants of the experiment were seniors, they were asked to complete an end-of-college survey and to make their academic transcripts available.

Amazingly, the one-hour intervention had a transformative effect for African Americans in the study but no effect for European Americans. Compared to African Americans in the control group and at Stanford as a whole, the grade deficit for African Americans in the adjustment-to-college group was cut in half. The percent of African Americans in the top 25 percent of their class tripled. The health deficit between African Americans and European Americans, measured by self-reported feelings of health and visits to a doctor's office, was eliminated entirely. All due to a one-hour intervention during their freshman year!

The daily surveys completed for a week after the intervention revealed what was taking place in the minds of the participants. For African Americans attending an elite university, the question "Do I belong?" was never far from their minds. It surfaced whenever something bad happened. The same was not true for European Americans, who interpreted bad things in other ways. The intervention allowed African-American freshmen to interpret the events of their daily lives as typical of incoming freshmen as opposed to unique to just them. As Walton and Cohen put it: "The intervention . . . untether[ed] their sense of belonging from daily hardship." In evolutionary terms, their symbotypes had been altered, a form of "symbo-therapy" conceptually similar to gene therapy.[26]

Let's take stock of our progress. The reason that I began this trio of chapters at the group level is to emphasize *the primacy of the small group as a unit of selection in human evolution*. I began this chapter on individuals by stressing that *individuals are products of social interactions*. Just because I can measure something about you, such as your grade point average, your physical health, or your mental health, doesn't mean that it's an essential property of you. Jim Coan's work shows that the human brain is wired to seamlessly integrate personal and social resources. Tony Biglan's work demonstrates that nurturance, or prosociality, is a master variable for human welfare. Now we have learned through the work of Walton and Cohen that success in college depends strongly on a sense of belonging, a feeling that we are part of "Us" and not "Them." If a one-hour intervention cut the minority academic deficit in half, what would a more comprehensive effort to increase a sense of belonging in marginalized students produce?

Also, it is the *perception* of belonging that counts. Walton and Cohen did nothing to alter the lives of the participants. They only altered the *worldview* of the participants, touching upon another major theme of this book: that the theory decides what can be observed. Returning to the metaphor of an adaptive landscape, it begins to make sense that the topography can be changed by the way you think about it. Viewed one way, you might be standing on top of a tiny hill, envying those atop taller hills, but unable to get there because every step that you take brings you down. Viewed another way, you can be magically placed on the slope of a mighty peak. As you chart your ascent, you might have to work around obstacles but every step can take you uphill. That's what ACT and related therapies can do for you. It doesn't necessarily require a lot of time and can improve your well-being in many ways, including but not restricted to your sense of belonging.

ACT works on both the variation and the selection part of your personal evolutionary trajectory. The variation part requires increasing your psychological and behavioral flexibility—to try out some new things rather than sticking to your old routines. The selection part requires you to reflect upon what is really important in your life so you can choose the behaviors that take you toward your valued goals.

One especially fast and effective form of ACT is called the Matrix, which is a space divided into four quadrants.[27] The lower half represents your symbotype, the world of thoughts and feelings inside your head. The upper half represents your phenotype, the actions that take place in the external world. The right half represents the thoughts, feelings, and actions that take you toward your valued goals. The left half represents the thoughts, feelings, and actions that take you away from your valued goals. Starting with the lower-right quadrant, spend a little time reflecting upon what is really important in your life. For example, think about a person who

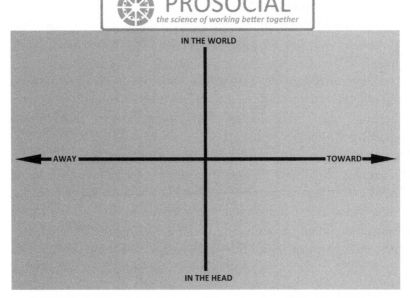

The Matrix is an exceptionally rapid form of ACT training. Prepare to evolve your own future . . .

means a lot to you. Why is this person important and what are your positive thoughts and feelings toward him or her? Draw the Matrix on a blank piece of paper and write some words in the lower-right quadrant if you like.

Now move over to the lower-left quadrant and spend some time thinking about negative thoughts and feelings that you have toward this person, much as you love them in other respects. Perhaps you become jealous when he or she pays attention to other people, become irritated by certain habits, find it difficult to forgive some past actions, or sometimes become angry to get your way. Moving into the upper-left quadrant, how do these negative thoughts and feelings manifest as actions that move you away from the relationship that you would like to have with this person? Finally, moving into the upper-right quadrant, what are the actions that will take you toward the relationship that you want? When you are done, imagine yourself at the crosshairs of the two lines. At any given moment, try to notice where your thoughts, feelings, and actions are located on the Matrix. Merely recognizing them as "toward" moves or "away" moves can help you to work toward your valued goals.

Congratulations! You have had your first ACT session, without charge and without needing to recline on a couch. You have increased your psychological flexibility (variation) and you are selecting the behaviors that take you toward your valued goals. You are beginning to consciously guide your evolution.[28]

Although the Matrix is a reflection that you undertook as an individual, it was also social (or could be) in three different ways. First, you were reflecting upon a relationship with a valued person. Even if I hadn't prompted you to do this, you probably would have reflected upon your social relationships in one way or another.

Second, Steve and his colleagues are increasingly imagining an individual person as a group of personas, with therapy a method of selecting for cooperative rather than conflictual interactions within ourselves. This is similar to imagining the genotype of an individual as a social group of genes.[29]

Third, the Matrix is proving to be a useful reflection for groups to increase their organizational flexibility and move toward valued collective goals. In the practical method for working with groups that I briefly described at the end of the last chapter, we begin with the Matrix as a warmup for reviewing the core design principles, as I will relate in more detail in the final chapter. In this fashion, we can consciously guide the evolution of our groups and multicellular societies in addition to our personal evolutionary trajectories.

Jim Coan, Tony Biglan, and Steve Hayes are at the peaks of their respective professions. At any time during their careers, they would have told you that they accepted Darwin's theory of evolution as a matter of course. Evolution figured here and there in their formal training, most notably in the behaviorist tradition of B. F. Skinner, who truly grasped the concept of individuals as evolving entities in their own right. But they didn't "get" the full application of an evolutionary worldview the way they do now. Only during the last decade or so have they first discovered, then adopted, and then contributed to a fully rounded approach to their subjects in which evolution appears at almost every juncture. They are at the forefront of completing the Darwinian revolution.

Now that we have considered groups and individuals, let's see what we can do at larger scales, and ultimately for the entire planet.

8

From Groups to Multicellular Society

The human population on earth is approximately 7.6 billion individuals. These people are divided into nearly two hundred nations that are highly variable in how well they govern their affairs. Governance at the transnational scale (for example, NATO, the United Nations, the European Union) is extremely weak and nations are competing with giant corporations for global influence. Together with our domestic plants and animals, we are rapidly crowding out the rest of nature. Our collective impact on the planet and its atmosphere has given rise to the term "anthropocene," a new geological age.[1] Scientific estimates of global warming have been too conservative. The earth is heating even faster than we feared it would. Yet many of our decision-makers are still in denial, and even those who accept the grim truth don't know what to do about it.

It is easy to conclude that the problems confronting us at the global scale are too large and complex to be solved, no matter what our theoretical perspective. But perceptions of size and complexity can be deceiving. Recall from past chapters that our bodies are composed of trillions of cells with billions being renewed daily, which serve as planets for a diverse ecosystem of thousands of other species. Yet, despite these huge numbers and complex interactions that we have only begun to fathom, our bodies, at least in health, function with a precision that puts a Swiss watch to shame.

Or consider that roughly ten thousand years ago, there were no large-scale human societies, only tribes of a few thousand individuals. Something happened in a wink of geological time that resulted in nations numbering in the millions and even billions of individuals. Transnational governance might be weak, but it would have been inconceivable two centuries ago. Compared to where we started ten thousand years ago, well-functioning governance at the scale of the whole earth seems just within reach.[2]

A single body is the gold standard for a well-functioning human society, as English words such as "corporation" that are derived from the Latin word for "body" (*corpus*) attest. Human societies have been metaphorically compared to single bodies since antiquity, from Aristotle's *Politics* to Hobbes's *Leviathan*, but only now has it become possible to place the metaphor on a firm scientific foundation. The first step toward viewing the whole planet as a single organism is to challenge the current orthodoxy and adopt the right theory.

THE INVISIBLE HAND IS DEAD

Arguably the most important concept in economics, from the origin of the profession to the present, is laissez-faire, which means "let it alone" in French.[3] The first known appearance of the phrase in print was in 1751, and the context was exactly the same as when the phrase is used today—to argue against government restrictions on trade. At that time, government was absolute monarchist rule and Christianity was the unquestioned worldview. It made sense to argue that there was a natural order, that governments were disrupting the natural order, and that the solution was to return to the natural order by letting things alone with respect to the economy.

Other milestones in the history of the laissez-faire concept were Bernard Mandeville's (1670–1733) *Fable of the Bees*, which fancifully compared human society to a beehive whose members are motivated entirely by greed,[4] and Adam Smith's (1723–1790) metaphor of an invisible hand, which guides selfish interests toward socially beneficial ends.[5] These expressions differed from each other. Conventional religious thinkers were scandalized by *The Fable of the Bees*, and Smith was not an admirer of Mandeville. However, all expressions of the laissez-faire concept relied upon the concept of a natural order—a system that works well as a system, with each element unknowingly doing its part. Without the concept of a natural order, there can be no justification for the prescription to let it alone.

The next milestone in the history of the laissez-faire concept was the French economist Léon Walras (1834–1910), who was inspired by the emerging science of physics to create a "physics of social behavior." It is easy to appreciate the allure of this goal. Isaac Newton and others had created mathematical equations that predicted such things as the motion of planets with astonishing accuracy. Wouldn't it be amazing to derive a comparable set of equations for economic systems? Walras and others succeeded at this goal, but only by making lots of simplifying assumptions about the preferences and abilities of individuals and the social environment in which economic transactions take place. Their work became the foundation for the school of economic thought that today is variously (and confusingly) called "orthodox," "neoclassical," and "neoliberal." The bundle of assumptions about human nature that is required for the mathematical equations to work is often called *Homo economicus*, as if it were a biological species.

All of this is pre-Darwinian. As we saw in chapter 4, Darwin's theory challenged the concept of a natural order, which made it profoundly disturbing to the Christian worldview. Even Darwin took a long time to realize this implication of his own theory. At first, he thought that his principle of natural selection could explain all of the design in nature that had been attributed to a creator. Gradually it dawned upon him that his theory could only explain functional design at the level of individual organisms. Morally praiseworthy traits such as honesty, bravery, and charity are selectively disadvantageous, compared to more self-serving traits within the same group. The same evolutionary process that created functional order at the level of the individual created functional disorder at the level of a social group. Darwin solved the problem to a degree by developing the concept of between-group selection, but that only notches functional design a little higher up the scale of nature—a war of groups against groups, rather than a war of individuals against individuals within groups.

Few people realize the degree to which the modern economics profession rests upon a pre-Darwinian foundation. The very concept of a "physics of social behavior" is misconceived, no matter how alluring it was at the time. No amount of tinkering with the assumptions of *Homo economicus* can save the mathematical edifice that seems to give the economics profession such authority. The first steps taken in a Darwinian direction by figures such as Joseph Schumpeter and Friedrich Hayek and their followers are only the beginning, and the inferences drawn from them are often mistaken. The field of behavioral economics, which challenges orthodox economics on empirical grounds, is still far from adopting Tinbergen's fully rounded four-question approach. In short, the concept of laissez-faire as we know it is dead as far as scientific justification is concerned, no matter how much it continues to influence political and economic policy.

LONG LIVE THE INVISIBLE HAND

In the days of absolute monarchs, the death of one and coronation of the next was sometimes announced with the words "The King Is Dead! Long Live the King!" In this spirit, I'm happy to herald a new concept of laissez-faire that can be justified by evolutionary theory. All that's needed is to replace the concept of a natural order with the concept of a unit of selection.[6]

If by laissez-faire we mean a society that functions well without members of the society having its interest in mind, then nature is replete with examples. Take the fabled bees that Mandeville chose to compare with human society. He was right that a bee colony functions well as a unit. He was also right that individual bees don't have the welfare of their colony in mind. After all, bees don't even have minds in the human sense of the word. But Mandeville was hilariously wrong to portray individual bees as like selfish human knaves. Instead, individual bees take part in the economy of the hive in the same way that individual neurons take place in the economy of the brain.

For example, when foragers return with their load of nectar and pollen, they are genetically programmed to dance in a way that conveys information about the location of the flower patch that they visited, with the length of the dance proportional to the quality of the patch. Other workers are programmed to pick a dancing bee at random and go where she went. The fact that some dances are longer than others introduces a statistical bias, directing more bees to the best patches without any bee making a decision. In this fashion, little snippets of behaviors by individual bees combine to make the colony function well as a unit. But there are any number of other behaviors that would not be good for the colony. Luckily, those negative behaviors were left on the cutting-room floor of natural selection. In other words, colonies whose members responded to the cues of their local environments in the right way survived and reproduced better than colonies whose members responded to the cues of their local environments in the wrong way. The result is an outstanding example of laissez-faire—a society that functions well without members of the society having its welfare in mind. It would never have happened without colony-level selection. Colony-level selection *is* the invisible hand that promotes the individual-level behaviors that benefit the common good rather than the much larger set of individual-level behaviors that would harm the common good.

The same story can be told for the genes, cells, and organs of multicellular organisms, which are the gold standard for a well-functioning society. To call a multicellular organism a society of lower-level elements is no longer metaphorical. It is literally the case that we are groups of groups of groups and that we qualify as organisms only because of our degree of functional organization, which evolved by between-organism selection. Also, a close look reveals that the ideal organism does not exist. Even the gold standard is marred with impurities that we can clearly recognize in human terms as disruptive self-serving behaviors at the level of cellular and genetic interactions, such as the example of cancer recounted in chapter 4.

RETHINKING REGULATION

As soon as we replace the concept of a natural order with the concept of a unit of selection, then the concept of regulation appears in a new light. Try to forget how politicians and economists talk about regulation and focus on how biologists talk about it. To regulate something means keeping it within certain boundaries—not too high or too low. The temperature in your home is regulated by a thermostat, a heating device such as a furnace, and a cooling device such as an air conditioner. The thermostat monitors the temperature and turns on the furnace or air conditioner, depending upon which is required to keep the temperature within bounds.

The number of processes in your body that need to be regulated to keep you alive is mind-boggling. Your temperature. The levels of glucose and CO_2 in your blood. Your blood pressure. Your sleep cycle. Your emotions. The list goes on and on. Each process has the equivalent of a thermostat, heater, and cooler. All of them have been provided to you by the process of between-organism selection, which favors the regulatory processes that work over processes that work less well or not at all.

In the bee colony, regulatory processes have been notched upward to include the social interactions among the bees in a colony. Social insect biologists even call these interactions "social physiology" to emphasize that they are orchestrated for the good of the colony as a whole. Take temperature regulation as an example. When the temperature in a beehive becomes too low, some workers start to vibrate their muscles to generate heat, just as we do when we shiver. When the temperature becomes too high, some workers fly away to collect water and return to spread it over the colony, in the same way that we sweat. The list of social regulatory processes in beehives and other social insect colonies goes on and on, like the list of physiological regulatory processes operating within each of us. A colony without regulation is a dead colony.

If human groups are to function like organisms, then they need to be regulated in the same way. What is the social physiology of small human groups that keeps our behaviors within bounds in the

service of collective goals? And how have these regulatory mechanisms been scaled up during the last ten thousand years of human
cultural evolution?

MY HALL OF SHAME

A trivial event in my own life illustrates the social physiology of
small human groups. One day when my kids were small I decided to
take them bowling. I seldom bowl, so I had forgotten that if someone in the lane next to you is about to bowl, it is customary to wait
until they have finished before you take your turn. After violating
this norm several times, an employee of the bowling alley discreetly
and politely informed me about the right way. There could not be
a more trivial incident, but my sense of shame was overpowering!
I was physically incapacitated and left the bowling alley with my
children as soon as I possibly could.

Here is another trivial episode from my personal Hall of Shame.
When one of my kids started taking karate lessons, I decided to
join her rather than just sitting on the sidelines waiting for the ses-
sions to end like the other parents did. I soon became fascinated
by the martial arts dojo. Several dozen people on the floor were
managed by our single sensei, like the conductor of an orchestra.
People paired up to perform the katas or to spar in a way that did
not result in physical injury. Each person functioned as either a student or a teacher, depending upon how their skill level compared to
that of their partner. Transitions from one activity to another were
accomplished swiftly with rituals such as bowing. In this fashion, an
extremely complex body of knowledge was being transmitted from
one person to another.

At the end of each session, everyone lined up in front of the sensei to learn if they had graduated to the next skill level, symbolized
by the coveted belt colors. If your name was called, you advanced
to face the sensei, bowed, and stood on his left side to receive your
new belt with a round of applause. When I earned my yellow belt,
I thoughtlessly stood on his right side. Once again, I was sweetly

shown the right way without any animosity, but I remember the shame burning in my ears to this day.

I know that I am not alone in my sensitivity to keeping in step with others in the dance of human life. It is part of what makes us human. We scarcely notice it most of the time because we are so practiced in our everyday lives. We have all earned our black belts in that regard. However, if you're like me, then as soon as you try to do something new within your own culture or visit another culture, you become acutely aware of the need to keep in step.

Nobody taught us to feel and act in this way. It is part of the psychological equipment issued to us by genetic evolution, based on thousands of generations of living in small groups.[7] Those who didn't care about abiding by norms did not prevail in the competition for survival. Either they failed within their groups because of a negative reaction on the part of other group members, or their entire groups failed for lack of a coordinated response to life's problems.

Life's problems are highly contingent. If you're an Inuit living in the Arctic, you need to coordinate your behavior in a different way than if you're an Mbuti living in the African equatorial jungle. Our basic psychological equipment is not learned and is roughly the same the world around, but it enables us to learn, regulate, and transmit the enormously complex body of knowledge that is required to survive and reproduce in any particular time and place. You can't have the learned knowledge without the genetically evolved psychological equipment.

To summarize, regulation, as biologists use the word, is needed to coordinate our lives in small human groups, just as much as it is needed for bodies and beehives. One more point needs to be made before we begin scaling up to larger human societies and ultimately the whole planet. Bodies and beehives provide outstanding examples of the invisible hand because they function well even though lower-level units don't have the whole's interest in mind. We know this for sure because genes, cells, and bees don't have minds in the human sense of the word. When we begin thinking about small human groups as units of selection in genetic evolution, we introduce the possibility that the lower-level units (individuals) *could* consciously behave to benefit the welfare of the higher-level unit (their group) because they do have minds. This is not *required*—people could be like genes, cells, and bees in this regard—but neither is it *prohibited*. It is easy to imagine a group whose members care only about the personal advantages that come with a good reputation and not the group as a whole. And such a group could work well if a good reputation requires behaving as a solid citizen. It is equally easy to imagine a group whose members do care about everyone's welfare and not just their own reputations. And this group could also work well. Which type of group is most likely to result from a process of between-group selection? This is a question about the *mechanisms* that underlie group-level performance (Tinbergen's mechanism question). If the mechanisms are functionally equivalent, what evolves might be a matter of *historical contingency*, like the different populations of *E. coli* that evolved to digest glucose in different ways (Tinbergen's history question). Once an evolutionary worldview has become part of your intuition, then this interplay of Tinbergen's four questions will be like second nature for you.

SCALING UP

Our capacity for rapid adaptation eventually led to a positive feedback loop in which the ability to produce food increased the size of human groups, which in turn enhanced the ability to produce more food. Paleoanthropology, archaeology, and recorded history provide a fossil record of this cultural evolutionary process. Unsurprisingly, the vast majority of historians have concentrated primarily on the history question and paid scant attention to Tinbergen's other three questions. An exception is the Harvard historian Daniel Lord Smail. In his book *On Deep History and the Brain*, Smail points out that most world histories begin about 4000 BC in the Middle East, which is suspiciously close to the time and location of the Garden of Eden according to biblical accounts.[8] Most historians are not young earth creationists, of course, so shouldn't they be extending their field of inquiry deeper in time and more widely in space? Shouldn't they also know something about the brain as a product of genetic evolution, which determines how people act at all times and places in a mechanistic sense?

The person who has advanced this agenda more than any other is not a historian but a biologist named Peter Turchin, son of the Russian physicist and dissident Valentine Turchin.[9] Peter immigrated to America with his father in 1978 and became a professor of ecology and evolutionary biology at the University of Connecticut, specializing in population dynamics. Many species in nature fluctuate in their numbers, with booms and busts that sometimes take the form of regular cycles and at other times are more chaotic. These fluctuations reflect complex interactions with other species in the community, along with exogenous factors such as climate. Using a combination of theoretical models and statistical tools for analyzing time series data, population biologists have become adept at explaining the booms and busts of species such as lemmings and bark beetles. Peter was among the best but when he reached midlife he decided that he needed a new challenge. He would apply his theoretical and statistical skills to the study of human history.

In a single stroke, Peter had isolated himself from almost everyone at his university and the wider world of academia. His colleagues in the Department of Ecology and Evolution never dreamt of studying human history and those in other departments who studied human history never dreamt of employing his theoretical and statistical toolkit. Indeed, so many attempts at grand historical narratives have come and gone (such as Marxism) that many historians have given up on the possibility of a unifying theoretical framework.

Undeterred, Peter announced the birth of a new field: cliodynamics, which combines the name of Clio, the muse of history in Greek mythology, with "dynamics," the study of how things change over time. That was fifteen years ago, and by now Peter's vision has fulfilled itself to a considerable degree. Thanks to him and a growing number of colleagues, we can begin to understand how cultural evolution favored ever-larger societies over the last ten thousand years, leading to the mega-societies of today. We can also zoom in on the United States of America as a boom-and-bust society, with periods of harmony that alternate with periods of discord over its 250-year history. Only when we understand history as part of evolution can we go beyond past efforts to improve national governance and work toward regulating our activities at a planetary scale.

THE LAST TEN THOUSAND YEARS

According to Peter's analysis, cultural evolution is a multilevel process much like genetic evolution. A learned behavior can spread through a population by benefitting individuals compared to other individuals in the same group, or by benefitting the whole group compared to other groups in the vicinity.[10] A learned behavior can even hop from head to head like a disease, at the expense of both individuals and groups, a possibility popularized by Richard Dawkins under the name "parasitic meme."[11]

The regulatory mechanisms that operate in small human groups are pretty good at weeding out learned behaviors that are parasitic or benefit some individuals at the expense of others within the same group. However, these mechanisms were not designed to work on a larger scale. The first agricultural societies therefore became despotic, ironically more like many animal societies than small-scale human societies. Despotic human societies are organized to benefit a small group of elites at the expense of everyone else in the group. As a basic matter of trade-offs, what it takes to remain in power as a despot is different from what it takes to function well as a group. You can't keep others under your thumb and then expect them to help you stay in power! Groups ruled by despots therefore tend to fare poorly in competition with more inclusive groups.

Between-group competition can take many forms, including but not restricted to direct warfare. Darwin stressed that nature is not always red in tooth and claw. A drought-resistant plant will outcompete a drought-susceptible plant in the desert, without the plants interacting with each other at all. The same goes for selection among human groups. A good example is the respect accorded to elders and deceased ancestors. Old people could easily be dominated by younger and stronger members of a group to the youngers' relative advantage (within-group selection), but such groups would not benefit from the knowledge possessed by older people (between-group selection). The between-group advantage could take the form of remembering the location of water holes during a drought or recalling a winning strategy in direct between-group competition.

While direct conflict is not the only form that group selection takes, it is one of the major forms whenever there is something to fight over—which was increasingly the case with the advent of agriculture. Multilevel selection does not make everything nice. It is an inescapable fact that our admirable ability to cooperate within groups is due very largely to violent competition among groups. This is not something we want for the future, but it is a true statement about the past.

Following Peter, let's drop in on two periods of human history, as revealed by written inscriptions on clay and stone tablets. The first is from Tiglath Pileser I, who ruled the Assyrian Empire from 1114 to 1076 BCE:

> Then I went into the country of Comukha, which was disobedient and withheld the tribute and offerings due to Ashur my Lord: I conquered the whole country of Comukha. I plundered their movables, their wealth, and their valuables. Their cities I burnt with fire, I destroyed and ruined . . . Their fighting men, in the middle of the forests, like wild beasts, I smote. Their carcasses filled the Tigris, and the tops of the mountains . . . The heavy yoke of my empire I imposed on them.

And here is one from Ashoka the Great, who ruled the Mauryan Empire in the region of current-day India and Pakistan between 268 and 239 BCE:

> Beloved-of-the-Gods speaks thus: This Dhamma edict was written twenty-six years after my coronation. My magistrates are working among the people, among many hundreds of thousands of people. The hearing of petitions

and the administration of justice has been left to them so
that they can do their duties confidently and fearlessly and
so that they can work for the welfare, happiness and benefit
of the people in the country. But they should remember
what causes happiness and sorrow, and being themselves
devoted to Dhamma, they should encourage the people in
the country to do the same, that they may attain happiness
in this world and the next.

Even adjusting for boastfulness in both emperors, what they
chose to boast about speaks volumes about a brutal despotic culture
on one hand and a more gentle inclusive culture on the other. What
accounted for the flowering of social justice (relatively speaking),
not just in the Mauryan Empire but throughout Eurasia during the
same period? According to Peter, it was not the cessation of warfare
but an increase in the *scale* of warfare.

The bullying tactics of despots who declare themselves to be
gods don't work above a certain scale. A relatively equitable society
that holds even its kings accountable is required to achieve the next
level of social organization; empires that span millions of square
kilometers and include tens of millions of people. The religions,
philosophies, and institutions that evolved during the so-called Axial
Age (roughly the eighth to the third centuries BCE) provided the
"social physiology" capable of holding such large societies together.
This includes the spread of Buddhism in India, Taoism and Confu-
cianism in China, and Christianity in the Roman Empire. They all
emerged as different solutions to the problem of internal coordination required for successful between-group competition at a massive scale.

Between-group selection does not always prevail over within-
group selection. A sufficiently fine-grained analysis of human history reveals both processes at work. Empires form in regions of
chronic between-group warfare, which acts as a crucible for the
cultural evolution of cooperative societies. Once one evolves that
can coordinate action at a larger scale than its rivals, it expands to
become an empire. Then cultural evolution takes place within the
empire, favoring self-serving behaviors and factionalism in myriad

forms. Eventually the empire collapses, like a cancer-ridden organism. According to Turchin, the centers of old empires become cultural wastelands for cooperation. New empires usually form at the edges of old empires, seldom at their centers. The increasing scale of society during the last ten thousand years has not been a smooth continuous curve. It has been a tug-of-war between levels of selection with a net gain for higher-level selection and many reversals along the way. The same tug-of-war operates in the present, with the evidence all around us once we know what to look for.

MULTILEVEL SELECTION AND CURRENT AFFAIRS

Ten thousand years of multilevel cultural evolution has led to the present moment, with roughly two hundred nations of various sizes, capable of regulating themselves for the common good to various degrees. Consider two small nations, Norway and Equatorial Guinea. Both are blessed with large oil reserves. Norway used its reserves to create the world's largest pension fund for the benefit of its citizens (worth over a trillion dollars). Equatorial Guinea used its reserves to enrich the president, his family, and a small circle of elites. For the rest of the population, the life expectancy is fifty-one years, and 77 percent have an income of less than two U.S. dollars per day.

In this and many other ways, it is obvious that Norway qualifies as a corporate unit (a society that functions like an organism) much better than Equatorial Guinea.[12] Books such as *Why Nations Fail*,[13] by the political historians Daron Acemoglu and James Robinson, and *The Spirit Level*,[14] by the social scientists Richard Wilkinson and Kate Pickett, measure and diagnose variation among nations in their capacity to provide a high quality of life for their citizens. Here is one of many graphs from *The Spirit Level*, which relates an index of health and social problems to the degree of income inequality. The Nordic countries anchor the "better" end of the distribution, along with Japan, Switzerland, and the Netherlands. The U.K., Portugal, and the USA anchor the "worse" end of the distribution with the USA an outlier in both income inequality

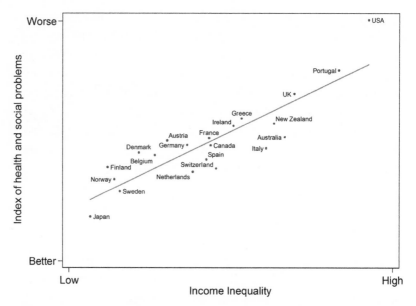

Modern nations vary greatly in how well they function as corporate units.

and health and social problems. According to the graph above and the more general diagnoses of these authors, the United States of America is doing a poor job regulating itself as a corporate unit.

These and related books have already been widely read and discussed in political and economic circles. The political economist Francis Fukuyama used the phrase "getting to Denmark" to pose the question of how any nation can work its way from the "worse" to the "better" end of graphs such as the one shown above.[15] Defenders of American governance in its current form have a long list of reasons why America isn't as badly off as the graph implies and why comparisons with the Nordic countries are inappropriate. They are so much smaller! They are more culturally homogeneous! Norway has oil (as if the USA doesn't)!

An analysis based on Tinbergen's four questions can add considerable value to these discussions, as all of the aforementioned authors are starting to appreciate. The history question acknowledges that cultural evolution, like all forms of evolution, can be highly path-dependent. Cultures don't just change in any direction. Different histories mean that different mechanisms for regulating social processes are likely to have evolved (the mechanism question) and these mechanisms are likely to be transmitted across generations in different ways (the development question). If "nation building" is possible at all, it must be much more attentive to historical, mechanistic, and developmental questions than it has been in the past.

Despite formidable difficulties in "getting to Denmark," Tinbergen's function question provides a blueprint for all nations to follow. One way or another, all nations must be structured to suppress disruptive self-serving behaviors by individuals and subgroups. They also must be structured to coordinate the right behaviors in diverse contexts, even when cheating is not a problem. If a nation isn't well regulated in the biological sense of the word, it will end up on the upper right side of graphs such as the one shown above.

Comparisons among nations are tricky because of their historical, mechanistic, and developmental differences. Perhaps the USA can never be like Norway or Denmark, but it doesn't need to be. It merely needs to be like the USA at previous points in its own history.

AGES OF UNITY AND DISCORD IN AMERICA

The analysis of variation *within* a given nation at different points of its history circumvents many of the problems associated with comparisons among nations. America wasn't always the land of inequality. It used to anchor the other end of the distribution, not once but twice during its 250-year history. In fact, the European colonization of the Americas provides a remarkable lesson in cultural multilevel selection, once we know what to look for.

As chronicled by Acemoglu and Robinson in *Why Nations Fail*, when the Spanish and Portuguese colonized Central and South America, they encountered societies that were already hierarchical and were able to become the new elite ruling class, governing for the benefit of themselves and not the population as a whole. This "extractive" social organization has continued to this day, illustrating just how entrenched extractive forms of governance can be. Most Central and South American nations function poorly as corporate units and always have. It is part of their cultural DNA.

The first British colonies in America, such as the Jamestown colony founded by the Virginia Company in 1607, also intended to rule over the Native Americans. When that didn't work, they tried to re-create a European feudal society by importing British laborers, who were made to live in barracks and work for rations; they were executed if they attempted to escape, much like slaves. But the British laborers had other options in America. Despite threat of death, they could and frequently did leave to become frontiersmen. By 1618, the colony was forced to adopt a more equitable social organization in order to survive. Each male settler was granted his own plot of land and all adult men had a say in the laws and institutions governing the colony. A very intense period of cultural evolution at the scale of small groups transformed an extractive social organization into an inclusive social organization, at least for adult white landowners. This experience was repeated in each of the thirteen colonies that coalesced to become the United States of America. That's why America grew to become a vibrant democracy (at least for white males), in contrast to the Central and South American nations.

Unfortunately, the checks and balances of the American Constitution did not entirely protect the fledgling nation from social processes that benefitted some people at the expense of others and the nation as a whole. To get the clearest view of America's swing toward inequality, it is best to turn from narrative historical accounts, such as the one offered by Acemoglu and Robinson in *Why Nations Fail*, to Peter Turchin's cliodynamic approach. The graph on page 189 is from Peter's newest book, *Ages of Discord: A Structural-Demographic Analysis of American History*. The solid line is an index of well-being, similar to the y-axis of the graph from *The Spirit Level* shown on page 186 (but with the poles reversed). It starts out high and rises to a peak in the 1830s. This is described as "the Era of Good Feelings"

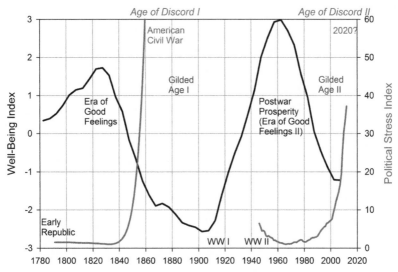

History as science. Well-being in America has swung like a pendulum during its history and is currently at a low point.

by historians and was immortalized by the French diplomat and social theorist Alexis de Tocqueville's *Democracy in America*, which he wrote based on his travels during 1831.

Then the index of well-being starts to decline. The America of 1900, dubbed "the Gilded Age" by historians, was no longer the America of the 1830s described by de Tocqueville. Notice that the Civil War took place at the halfway point of the decline. Oddly, this cataclysmic event in American history neither caused the decline nor seemed to have had much of an effect on the index of well-being.

Like a pendulum moved by hidden forces, the index of well-being begins to rise again, reaching a peak during the 1960s. The Great Depression, another convulsive event in American history, occurs partway up this slope. Like the Civil War, it seems neither to have caused nor to have had much of an effect on the pendulum swing. The next decline begins in the 1970s. The election of Ronald Reagan, which is regarded as a watershed in American political history, occurs partway down this slope, more like a symptom than a cause.

Graphs such as the two that I have presented are so common that we tend to either accept or reject them on faith, trusting or distrusting the experts who made them, without making our own examination. It is therefore worth spending a bit more time describing Peter's methods as a master evolutionary scientist at work. The kind of census data that Wilkinson and Pickett use in *The Spirit Level* doesn't extend very far back in time. To create a well-being curve for the entire span of American history, Peter used four proxies that could be assembled from historical records. The first is an index of relative wages, or income inequality. The second is physical stature (height at adulthood). The third is average age of marriage. The fourth is average life expectancy. Three of the four proxies are hard biological measures of well-being. If you didn't have enough food to grow as tall as you might have, if you had to wait longer to get married than you might have, and if you flat-out died sooner than you might have, then your well-being was comparatively low.

Each proxy for well-being is based on historical information of variable quality. Even though the data is noisy, when you chart the four proxies separately on a graph, they line up fairly well with each other. When you estimate something in four different ways using different sources of information and they all line up, something is going on that isn't likely to be attributed to measurement error. The more you examine Peter's methods of gathering historical time series data, the more you trust that the pendulum swings shown in the figure are real.

That's only half of what Peter accomplishes in *Ages of Discord*. The other half is a model of the social forces that cause the pendulum of well-being to swing. The model includes four major compartments and their interrelations: (1) The State (Size, Revenues, Expenditures, Debt Legitimacy); (2) The Population (Numbers, Age Structure, Urbanization, Incomes, Social Optimism); (3) The Elites (Numbers, Composition, Incomes and Wealth, Conspicuous Consumption, Social Cooperation Norms, Intra-elite Competition/Conflict); and (4) Forces of Instability (Radical Ideologies, Terrorism and Riots, Revolution and Civil War). In addition to his index of well-being, Peter has estimated all of these other variables with historical data, which he uses to build computer simulations—virtual Americas to compare with the real America.

If all of this sounds dauntingly complex, it is no more so than modeling the pendulum swings of bark beetle populations, forest fires, and epidemics, which was Peter's previous line of work. In other words, the complexity does not prevent us from making solid scientific progress. For example, models of disease transmission often divide the host population into three compartments: susceptible individuals, infected individuals, and individuals that have recovered and become immune. Peter uses the same models, virtually unchanged, to examine the cultural transmission of radical political ideologies.

The gray lines in Peter's graph are a political stress index, which is based on the voting records of the two major political parties in the United States Congress (currently the Democratic and Republican parties). The ideological differences between the two parties were minor up to the 1840s and then spiked upward just prior to the Civil War. Alarmingly, the same trend is occurring today. We seem to be on the brink of a new Age of Discord. It probably won't take the form of a civil war, but it might take the form of riots, terrorist acts, and social chaos. Worse, other nations seem to be on the brink of their own Ages of Discord. In the tug-of-war between levels of selection, disruptive within-group competition appears to be winning.

Thankfully, the situation is not hopeless. Human history is not like a mechanical pendulum. It responds to human agency. By taking the right kind of collective action, we can pull back from the brink. This is not just a conjecture about the future; it is a statement that we can make about our own past.

Let's return to the last nadir of well-being in American history, the early 1900s. Income inequality was at an all-time high. The rich were never richer and the poor were working for starvation wages. The economy was collapsing. Riots and terrorist events were on the rise and communism was threatening to upset the entire world order. The situation was so dire that some of the elites of America began to formulate policy with the welfare of the whole nation in mind, even at their own relative expense. Put another way, they realized that if they didn't subordinate their own well-being to the well-being of the nation, they might well go down with the ship. Their decision was like the decision of the Virginia Company that its tiny colony on the banks of the James River must become more inclusive to survive.

Thanks to this collective decision, America pulled back from the brink and the well-being of the average American started to rise. They didn't just bring home a larger paycheck—they grew taller, married earlier, and lived longer. What could have become the Era of Total Collapse turned into the New Deal, which Peter has dubbed the Second Era of Good Feelings, to stress its similarity to the previous one.

John D. Rockefeller is one of the corporate giants who had a change of heart. Previously a believer that his money was a sign from God and that industrial paternalism (power concentrated in the hands of people like himself) was best for the nation, he made this statement in 1919: "Representation is a principle which is fundamentally just and vital to the successful conduct of industry . . . Surely it is not consistent for us as Americans to demand democracy in government and practice autocracy in industry . . . With the developments of industry what they are today there is sure to come a progressive evolution from autocratic single control, whether by capital, labor, or the state, to democratic cooperative control by all three."

And so there was. Labor unions became strong, the state played an active role in social welfare and industrial development, and the tax rate on the top income bracket rose to a level comparable to that in the Nordic countries today. These changes were not forced upon the elites—they were accomplished by the elites based on a political philosophy that stressed the welfare of the whole nation, or the whole organism, as necessary for their own long-term welfare. But alas, people have a way of forgetting the lessons of history, especially when they can profit by their forgetfulness over the short term. Selfish interests and ideologies that justified individual and corporate greed as good for the whole group began to spread, leading to our decline in well-being and increase in political discord to the present moment, as Peter's graph shows.

THE FINAL RUNG

If ever there was a case of a wrong theory blinding us to reality, it is the current "greed is good" ethos of orthodox economic theory. However, you can't discard an old theory without adopting a new one. An evolutionary worldview provides a robust alternative that is remarkably simple to envision. We merely need to replace the pre-Darwinian concept of a natural order with the Darwinian concept of a unit of selection. Multilevel selection theory provides a framework for understanding the design of small groups, the increase in the scale of society over the last ten thousand years, and the current frontiers of human social organization, which can either advance to achieve a global scale or disintegrate into smaller corporate units that make life miserable for each other.

One conclusion is that for any group to function as a corporate unit, the potential for disruptive self-serving behaviors within the group must be suppressed. This conclusion is so basic that it applies to nonhuman species in addition to human groups of any size. It is remarkable that the core problem confronting America in the early twentieth century and today is the same problem that confronted the tiny Jamestown colony in the early 1600s: an extractive social organization that allows some to gain at the expense of others and the group as a whole. This is the human societal equivalent of cancer.

It is important to stress that disruptive self-serving behaviors need to be judged by actions, not intentions. Consider Alan Greenspan, the economist who served as the U.S. Federal Reserve Board chairman from 1987 to 2006. By all accounts he was a well-intentioned man who thought he was improving the welfare of the nation and the world, but he was a major player in the decisions that led to the decline of well-being in America during this period. A well-intentioned person blinded by the wrong theory is little better than a person driven by selfish impulses. As the novelist and existential philosopher Albert Camus wrote: "The evil of the world comes almost always from ignorance, and goodwill can cause as much damage as ill-will if it is not enlightened . . . There is no true goodness or fine love without the greatest possible degree of clear-sightedness."

Another conclusion is that for any group to function as a corporate unit, it must be well regulated in the biological sense of the word. Countless processes must be kept within bounds to meet the challenges of survival and reproduction. What is true for a multicellular organism or a social insect colony is also true for a human society. Whenever I hear talk of regulations as categorically bad, I feel like shouting "An unregulated organism is a dead organism!"

Yet it is also absurd to think that all regulations are good. Regulations are like mutations—for every one that works well for a group, there are many that work poorly. Also, it is a mistake to associate regulation exclusively with government controls and centralized planning. Most regulatory systems in the biological world are decentralized and self-organizing. Much of the regulation in small-scale human groups takes place spontaneously and beneath conscious awareness, thanks to our genetically evolved psychological mechanisms. The biological concept of regulation does not map onto any current political ideology. That's why an evolutionary worldview is in a position to provide new solutions that can appeal across the political spectrum.

A third conclusion follows directly from multilevel selection theory. Adaptation at any level of a multi-tier hierarchy requires a process of selection *at that level*. The whole cannot be optimized by separately optimizing the parts. That's why the concept of the invisible hand in orthodox economic theory, which pretends that the pursuit of lower-level self-interest robustly benefits the common good, is so deeply flawed.

As an example, consider so-called "smart city" movements that are taking place around the world. To say that a city can be smart is to say that it can have something analogous to a nervous system and brain, which receives and processes information in a way that leads to effective action. This can include a technological component, such as electronic sensors that monitor traffic flow, but it also can include a human component, such as 311, a three-digit number that residents of a city can call to communicate a problem such as a pothole, a fallen tree, or a failure to pick up garbage.[16] The use of the number originated as a "cultural mutation" in Baltimore, Maryland, in the 1990s to process calls that were being inappropriately made to 911, which is designed to handle urgent matters. Soon the potential of 311 to turn residents into the "eyes and ears" of a city began to become apparent, with cities such as Boston taking the lead. When Hurricane Irene struck the East Coast of the United States in 2011, the city of Boston received 1,045 reports of "tree emergencies" during a forty-eight-hour period through its 311 system, enabling it to decide how to deploy its maintenance crews without needing to gather the information in other ways. Currently the city receives about 175,000 calls to 311 annually, and similar systems have been implemented in over four hundred American cities.

The metaphorical description of 311 as the "eyes and ears of the city," along with catchphrases such as "the pulse of the city," indicates the intuitive appeal of thinking of a city as a single organism with a "social physiology" that receives and processes information in a way that leads to effective action. These processes seem "automatic," "effortless," and "self-organizing" in single organisms. We see and hear without needing to think about it—but only thanks to enormously complex mechanisms that evolved by natural selection at the level of single organisms. If something similar is going to evolve at the scale of a city, it will need to be selected as a whole system. Systems engineers have known this for a long time, but only recently have they started thinking of what they do as a highly managed form of cultural group selection.[17]

It is useful to make a distinction between "natural" and "artificial" cultural group selection, similar to the distinction between natural and artificial selection in the biological sciences. The shape and coloration of a moth that exactly resembles a leaf is a product of natural selection, based on the removal of more conspicuous moths by predators. The shape and coloration of the flowers in your garden are a product of artificial selection, based on the removal of plants with less showy flowers by humans. The same basic principles of evolution are at play in both cases. Also, the distinction between natural and artificial selection is not always clear-cut. In many species of animals, members of one sex (usually but not always the males) are highly ornamented because they were selected as mates by members of the opposite sex, much like human gardeners selecting showy flowers. The domestication of dogs probably began with the natural selection of wolves that began to hang around human habitations, not the deliberate selection of these wolves by people.

In human cultural evolution, the line between artificial and natural selection is even more blurred. Human cultural change is intentionally guided to some extent but it is also a product of many inadvertent social experiments, some that hang together and others that fall apart. The Protestant Reformation provides an excellent example. Reformers such as Martin Luther and John Calvin were very deliberative in the construction of their theologies and social ordinances, but the consequences of their intentional efforts were highly variable, based on unforeseen consequences. Calvinism in Geneva was not the sole product of Calvin but a complex interaction between Calvin's efforts and many other people and events. The fact that this complex interaction resulted in a city that functioned well, compared to the efforts of other reformers such as Huldrych Zwingli in the city of Zurich, would have been impossible to predict beforehand. Intentional actions are often converted into chance actions when they collide with each other!

The most important conclusion for our purposes is that cultural group selection must become more intentional and deliberative than ever before. We can't wait decades and centuries for our inadvertently successful social experiments to spread. We must learn to function in two capacities: as *designers* of social systems and as *participants* in the social systems that we design. As designers, we must have the welfare of the whole system in mind, such as the welfare of a city in the construction of a 311 telephone system. The invisible hand concept does not apply. As participants, we can only have our own local concerns in mind, such as punching three digits into our smartphones when there is a problem in our neighborhood. The invisible hand concept does apply. In short, system-level selection *is* the invisible hand that causes local actions to benefit the common good. The invisible hand must be *constructed*, which is a contradiction of terms according to the concept of laissez-faire.

Finally, even if we succeeded at creating smart cities or smart nations, that would not solve the problems of coordination at larger scales. The conclusion is inescapable: To solve problems at the planetary scale, we must formulate policies with the welfare of the whole planet in mind. "Me first," "Our corporation first," and "Our nation first" are not good enough. It must be "Our world first." The lower-level units don't disappear. On the contrary, their welfare is paramount, but in the process of thriving they must contribute to rather than undermine the higher-level common good. A multicellular organism can't be healthy without its cells and organ systems being healthy—but there is still a difference between a healthy normal cell and a healthy cancer cell. Even a nation such as Norway, which works so well at the national scale, must regulate its behavior (such as the global investment strategy of its pension fund) to avoid becoming part of the problem at the planetary scale.[18]

The challenges of getting all nations to function as well as Norway (or Denmark) are daunting, not to speak of ascending the final rung to become a worldwide corporate unit, but at least we have the right theory to guide us.

9

Adapting to Change

Change is the mantra of modern life. Gone are the days when every generation was much like the one that preceded it. Now, every decade seems to be transformational. The changes are driven by our own activities in pursuit of what we think we desire, but like the character of a folktale who is granted a wish, we often end up regretting what we wished for. Sometimes we don't even realize that we are the agents of our own misfortune and remain permanently trapped in a maze of unforeseen consequences.

It is ironic that in the midst of all this change and the use of keywords such as "evolve" and "adapt" in the vernacular, so little effort has been made to consult evolutionary theory. In this chapter I will show that for a group to be *adaptable* to changing environments is different from being *well adapted* to a current environment. I will provide some examples of adaptable groups drawn from the literature on businesses. Between-group competition in the business world is so intense that it acts as a crucible for the cultural evolution of effective change methods, since adaptable firms will survive longer and replicate their social organizations more than firms that are incapable of change. This is what economists such as Joseph Schumpeter and Friedrich Hayek meant by phrases such as "creative destruction" and "spontaneous order." They took a few steps toward an evolutionary worldview, perhaps as many steps as could be taken during their time, but modern evolutionary theory provides a much more powerful and diverse toolkit for understanding the cultural evolution of effective change methods in the business world and how they, in turn, can be adapted to effectively manage change in other walks of life.

Before I review change methods that work, let's clear the deck of change methods that won't work.

WHAT WON'T WORK

One thing that won't work is laissez-faire—the idea that our societies will work best if we leave them alone. If there is anything that evolution teaches us, it is that the pursuit of lower-level self-interest does not automatically benefit the common good. As we have seen in previous chapters, evolution frequently results in behaviors that are good for me but not you, us but not them, or all of us today without regard for future generations. If we want to manage our societies for the common good, ultimately at the global scale, then it is up to us to navigate toward that goal. There is no natural order that will do it for us.

Yet another thing that won't work is centralized planning, in which a group of experts decides what needs to be done and proceeds to execute their grand plan. The reason that centralized planning seldom works is because the world is too complex to be understood by anyone.[1] No matter how smart the decision-makers or how well informed by theory, there is always the likelihood of unforeseen consequences. This is true not only at the scale of nations, where efforts at centralized planning have failed repeatedly, but also at the scale of business firms, where command-and-control change efforts have a very poor track record.[2]

WHAT CAN WORK

There is something between centralized planning and laissez-faire that can work: *a managed process of variation and selection*. First there must be a target that we are aiming to hit. For a business, this might be profitability, social responsibility, or the development of a new product line. At the planetary scale, it might be a robust global economy or the reduction of greenhouse gases. These targets become the criteria for selection, similar to an animal breeder such as William Muir selecting for egg productivity in hens or the target of creating a smart city with a 311 system for reporting dysfunctions.

Then there must be variation, which exists in two forms: planned and unplanned. We can deliberately formulate alternative strategies for reaching the target goals, comparing them to each other and to current practices using rigorous statistical methods. This is how I evaluated the Regents Academy, Ron Prinz evaluated Triple P, and Steve Hayes evaluated his self-help book. Science itself can be regarded as a carefully managed variation-and-selection process of this sort, where the target of selection is the advancement of objective knowledge.[3]

In addition, unplanned variation exists all around us. Every firm has its own way of doing things, every state has its own health care plan, and nations vary widely on a scale of well-being, as we saw in the previous chapter. Calling this variation "unplanned" isn't quite right because a lot of thought might have gone into the practices of each group. Nevertheless, the groups are likely to differ in so many ways that identifying the differences most relevant to the target of selection can be difficult. If your firm is more profitable than mine, what exactly should I copy? On the other hand, unplanned variation can include successful practices that are beyond the imagination of those making planned comparisons. It is therefore wisest to include both planned and unplanned variation in the variation-and-selection process.

In chapter 5, I introduced the important concept of an evolutionary process built by another evolutionary process.[4] An evolutionary process such as the immune system, individual learning, and transgenerational cultural change is itself a product of genetic evolution. Actually, genetic evolution by itself illustrates the same concept, since the current mechanisms of genetic inheritance are far more sophisticated than their ancient precursors.[5] The mechanisms

A Toyota assembly plant is a symphony between people and machines that produces a car every minute.

of human cultural evolution have also changed over the centuries, especially with the advent of new technologies such as writing, the printing press, and computers, which enable vastly more information to be transmitted across generations than before.[6] A detailed analysis of the archaeological, anthropological, and historical records will probably document the process of more adaptable societies replacing less adaptable societies over the last ten thousand years. Business firms are a type of society, so the same approach can potentially explain the cultural evolution of successful change methods in the business world. Here are three examples.

LET'S GO PLACES

If you're a businessperson, it's no secret that Toyota is a shining example of a corporation that works well and adapts to the ever-changing environment of automobile manufacturing. Founded in 1937, it has become the second-largest automotive corporation in the world (following the German Volkswagen group) and is the world leader in hybrid electric vehicle sales. An entire genre of business books is devoted to analyzing its success. Yet emulating the success of Toyota is easier said than done. An automobile manufacturing plant rivals a multicellular organism or a social insect colony in the complexity of its design. Consider the queen of a termite colony, who is fed and protected by an army of workers and lays an egg from her hugely distended abdomen every few seconds. Now consider Toyota's manufacturing plant in Georgetown, Kentucky, which also houses an army of workers and spits out an automobile at a rate of about one per minute. Each automobile is so well designed that it can be driven several hundred thousand miles, which is roughly the distance from the earth to the moon. Take a visitor's tour of the plant and you will witness what is sometimes described as a symphony between people and machines, as the cars move along a conveyor belt in a system pioneered by Henry Ford in the early twentieth century.

How is efficiency and productivity monitored in such a complex system and how are improvements made? Although business leaders complain about centralized planning at the level of national governments, they often employ a "command-and-control" form of social organization in their own companies. The management runs the company in a top-down manner by gathering data on performance and issuing orders that staff are expected to execute. If performance doesn't improve, then the unsuccessful managers are fired and new ones are hired. This is a form of cultural evolution that works to a degree, just as the failure of whole businesses and their replacement by other businesses works to a degree, but the success of command-and-control is limited for a single business in the same way that it is limited for government. Human social organizations are too complex for anyone to know how they work, and the kind of performance data that is usually collected is too aggregated to be very useful. Knowing that a given unit failed to meet its productivity quota or that a given part has an unacceptable proportion of defects doesn't tell you what needs to be done to improve the situation.

Here's an example of the kind of complexity that an automobile assembly plant is confronted with on a daily basis. Let's say that at a particular point on the assembly line, the operators need to keep a number of parts within easy reach. This means that the parts need to be in relatively small bins and need to be replenished as they are used up. The parts are brought to the plant in large lots, so someone has to divide them into small lots and physically transport them to the assembly line at frequent time intervals. Less effort for the assembly line worker requires more effort for the receiving and distribution departments. What's the best way to manage this trade-off to maximize overall productivity? Countless trade-offs like this must be managed, all interacting with each other, similar to trade-offs in the physiology of a multicellular organism or the social physiology of an insect colony.

Toyota has a way of continuously improving its operation that dates back to its origin as a company that made weaving looms. Here is how Sakichi Toyoda, founder of the Toyoda Automatic Loom Works, responded when the design plans for one of his looms was stolen.

> Certainly the thieves may be able to follow the design plans and produce a loom. But we are modifying and improving our looms every day. So by the time the thieves have produced a loom from the plans they stole, we will have already advanced beyond that point. And because they do not have the expertise gained from the failures it took to produce the original, they will waste a great deal more time than us as they move to improve their loom. We need not be concerned about what happened. We need only to continue, as always, making our improvements.[7]

A key phrase in this passage is "experience gained from failures." Most people think about failure as something to be avoided. In most businesses, people are promoted for their successes, which provides a strong incentive to avoid taking risks and to conceal failures when they occur. From an evolutionary perspective, however, failures are the current frontier of adaptation. Every failure provides an opportunity for a variation-and-selection process to go to work to improve the efficiency of the whole operation.

In the past, Toyota assembly plants had cords called "andons" hanging down from the ceiling that operators were instructed to pull whenever an inefficiency occurred at their station. These cords, which have been replaced by more sophisticated monitoring equipment in modern Toyota plants, functioned like the pain receptors of a multicellular organism or the alarm signaling systems of social insect colonies. Just as poking a hole in a termite mound results in a swarm of activity to repair the damage, pulling the andon resulted in a swarm of activity to solve the problem. Plant managers had their offices located on the shop floor, rather than in a separate location, so that they could be directly involved in working with the lower-level employees who encountered the problem. Since the assembly line workers are closest to the problem, their knowledge is often essential for coming up with a solution, so that decision-making becomes both a bottom-up and a top-down process focused on solving problems at a very fine level of detail, rather than managers issuing directives on the basis of aggregated data such as quarterly reports.

When a tiny change is made in a complex system, the consequences can be magnified by the interactions among elements of the system. This is why the weather is so unpredictable, which gave rise to the term "butterfly effect"—a butterfly flapping its wings in Africa can result in a hurricane in Honduras.[8] In an automobile assembly plant, a small improvement in one part of the system can easily disrupt other parts of the system. No one is smart enough to anticipate all of the indirect effects, so it is necessary to experiment. Toyota has learned from experience to implement only one change at a time and to monitor the effect on the whole system before adopting the change. Even making a few changes at the same time would result in interactions that are too difficult to track.

With a monitoring and improvement system in place, Toyota sets its production quotas so that failures will occur. Here is how the complex systems analyst Mike Rother describes this distinctively positive attitude toward failure in his book *Toyota Kata*:[9]

"No Problem" = A Problem

At a Toyota assembly plant, I was once told that the normal number of andon pulls is typically around 1,000 per shift. Each pull is an operator calling for assistance from their team leader because the operator is experiencing a problem; a cross-threaded bolt here, a task that took a little too long there. Naturally the number of andon pulls per shift varies, and I once heard of it dropping to only 700 pulls/shift. When I ask non-Toyota managers what they would do in this situation, I often get the answer, "We would celebrate the improvement."

According to my source, what actually happened when the number of andon pulls dropped from 1,000 to 700 per shift is that the Toyota plant's president called an all-employee meeting and said, "The drop in andon pulls can only mean two things. One is that we are having problems but you are not calling for help. I want to remind you of your responsibility to pull the andon cord for every problem. The other possibility is that we are actually experiencing fewer problems. But there is still waste in our system and we are staffed to handle 1,000 pulls per shift. So I am asking group leaders to monitor the situation and reduce inventory buffers where necessary so we can get back to 1,000 andon pulls per shift.

Inventory buffers are stocks of parts that workers can draw upon when the parts are not coming down the assembly line. They increase productivity over the short term but disguise inefficiencies in the process. Reducing the inventory buffers causes the workers to stand idle when the parts aren't coming down the assembly line. This decreases productivity in the short term but increases it over the long term by revealing and eliminating inefficiencies in the whole system.

Toyota's systemic approach to failure as the frontier of improvement is reflected in its policies toward both workers and managers. In companies governed by command-and-control, success and failure is attributed to individuals. If goals aren't being reached, then the response is to fire and hire until they are. Toyota's default assumption is that its employees are talented and hardworking. When problems occur, it is the *system* that needs improving and workers and managers need to work together to improve the system. Firing an individual for a systemic problem does not solve the problem. Toyota therefore retains its employees and recruits leaders from its own ranks more than other automobile manufacturers. It also has a system for perpetuating its culture that involves every employee having a mentor.

So far I have described how Toyota responds to failures in the same way that a multicellular organism responds to pain or a social insect colony responds to threat. In addition, Toyota has an impressive ability to move toward positive long-term goals in the same way that an individual person aspires to do more than merely react to pain. This requires first having a long-term aspirational goal (the target of selection) and then formulating concrete ways to move toward it in a step-by-step fashion. If the target goal had been to maximize quarterly earnings, it would have led to a different set of practices. Toyota's change method bears an intriguing similarity to Acceptance and Commitment Therapy (ACT), which empowers individuals to mindfully work toward their valued long-term goals.

Toyota is exceptionally good at improving its operations and adapting to its ever-changing environment—not by adopting a laissez-faire policy, not by centralized planning, but by a sophisticated process of variation and selection. Its adaptability is largely responsible for its dominance in the automotive sector. Companies that are less adaptable either go under or remain in the marketplace by copying success. This is the process of creative destruction imagined by economists such as Joseph Schumpeter, but a modern evolutionary perspective paints a more complicated picture.

Let's think about the Toyoda Automatic Loom Works as a mutant firm. However its founder came by the method of continuous improvement, it made his firm more adaptable than competing firms. The Toyoda firm captured the loom market, an example of firm-level selection. Then the same method spread to other business sectors, such as automobile manufacturing.

Competing firms were quick to notice the success of Toyota and eager to emulate its methods. Cultural evolution often takes the form of groups copying the best practices of other groups, rather than going under and being replaced by them. But here we encounter a major complication. Firms that attempt to copy Toyota often end up copying its *current practices*, like the thieves who stole the design plan for one of the looms, without copying *the evolutionary process that gave rise to the current practices*. This is why efforts to emulate Toyota so often fail.

Years were required for complex systems analysts such as Mike Rother to figure out the processes that make Toyota work, in part because even Toyota executives couldn't articulate them. Cultural evolution often results in practices that work without anyone knowing why they work. Rother and his associates succeeded in part because they were scientifically oriented (a rarity in the field of business and management), had a deep appreciation of complex systems, and had a workable knowledge of other fields such as evolution and psychology. Thanks to their efforts, other business firms can do a better job emulating Toyota because they can copy the evolutionary process rather than merely imitating current practices. Finally, Rother is fully aware that the variation-and-selection process built by Toyota can be applied to many other business sectors and in non-business contexts, including individuals adapting to their everyday lives.[10]

That's the good news. The bad news is that roughly a century has elapsed since this cultural mutation first arose in Japan early in the twentieth century. While it has spread, it has not spread fast or far enough. I'm betting that less than 5 percent of the readers of this book are familiar with the material about Toyota that I have presented. In the meantime, elements of Toyota's "social physiology" are becoming outdated. In the past, Toyota has perpetuated its culture by having every employee mentored by a more experienced employee (what Rother calls the Coaching Kata). That system won't work when a new factory has to be built in a foreign country and only a small fraction of the workforce is experienced. Whether Toyota can adapt to a new transnational and multicultural environment remains to be determined.

Clearly, something more is required for evolutionary processes exemplified by Toyota to become widespread in all walks of life. Another cultural mutation that took place in the business world can add to the insights that we have gleaned from Toyota.

RAPID RESULTS

The biological world is full of examples of convergent evolution—species that resemble each other, not because they are historically related, but because they have experienced the same selection pressures. Convergent adaptations resemble each other functionally (for example, the hard shells of turtles and armadillos or the eye of the octopus compared to the vertebrate eye) but usually differ in their mechanisms and development by virtue of their different historical origins, illustrating the need to ask all four of Tinbergen's questions in conjunction with each other.

Examples of convergent cultural evolution can also be found in the business world, including a change method called "rapid results" that converges upon Toyota's adaptability in functional terms, without any historical connection, using different procedures and maintaining cultural continuity in different ways.[11] The rapid results method was the brainchild of Robert H. Schaffer, a business consultant who observed how hard and well people work together in emergency situations. Not only do they pull off miracles, but they experience intense pleasure doing it, despite the emergency situation being dysphoric in other respects. This by itself can be understood from an evolutionary perspective, since life-and-death situations requiring group solidarity occurred regularly throughout our evolutionary history. Our minds are prepared for them.[12]

The specific event that inspired these thoughts was a wildcat strike that took place at a New Jersey oil refinery. About 450 supervisors, managers, and engineers were forced to run an operation that normally employed a workforce of 3,000, not just for a few days, but for four months—and they did it well.

Schaffer wondered if this kind of peak performance could be elicited under normal working conditions. He experimented with creating "emergency" situations by challenging small teams within a company to accomplish daunting goals in a short amount of time, such as doubling the number of customers for a new product line in 100 days. He discovered that the teams indeed leapt into action and displayed the same kind of zest as in an emergency situation. Hence, the concept of a "rapid results cycle" was born.

The rapid results method produced additional benefits. It turned out that when lower-ranking employees most closely associated with a given challenge (such as increasing sales or customer satisfaction) were tasked with solving the problem, they came up with better solutions than what top managers or outside consultants might suggest. This was because they were in a better position to understand the nature of the problem and devise workable solutions (note the convergence with Toyota's practices). The "emergency" atmosphere also allowed bureaucratic rules to be relaxed. And group members were able to bask in their success rather than watch all the credit going to their bosses. Upper-level managers sometimes had difficulty relinquishing control and trusting the bottom-up process, but the results spoke for themselves.

Another benefit of the rapid results method was more subtle. As we have seen with Toyota, large companies are highly complex systems with many parts that must work in a coordinated fashion to function as a whole. The whole system cannot be optimized by trying to separately optimize each part, and nobody is smart enough to anticipate all of the interactions among the parts. Accomplishing positive change in a large company is therefore a formidable task, and top-down efforts are more likely to fail than succeed. The changes produced by rapid results cycles were small enough to be integrated with the larger business operation, like the small changes that Toyota makes in its assembly line operations. Far from nibbling at the edges of fundamental change, rapid results cycles could become an engine of fundamental change in an incremental fashion, producing short-term benefits along the way and without requiring expensive outside consultants. This combination of benefits might seem too good to be true, but it has been documented repeatedly and some major corporations have adopted rapid results cycles as their main change engine. It is important to stress that this requires both a bottom-up process (the rapid results teams) and a top-down process (a strategy for employing the teams in a way that results in a long-term systemic goal). Either process by itself would be inadequate.

Just as Toyota's adaptable social organization spread from one business manufacturing sector (looms) to another (cars), the rapid results method spread from the business world to the seemingly very different world of international aid through the creation of a nonprofit organization called the Rapid Results Institute.[13] As one example, previous efforts to persuade women to visit family planning clinics in Madagascar resulted in gains of only a few percentage points over a fifteen-year period. Rapid results teams composed of local women committed to the seemingly impossible goal of increasing the percentage by 30 percent in 100 days. Not only did the groups meet their goals, but some achieved gains as high as 500 percent. One can well imagine the women springing into action in a way that they would not have done before. The rapid results method is currently being employed in over a dozen developing countries on problems as diverse as child malnutrition, HIV/AIDS, and corruption.

Like Toyota, the rapid results method provides an example of cultural evolution in action. Its origin was serendipitous and idiosyncratic, much like a biological mutation. Nevertheless, it spread on the basis of its success. In functional terms, it converges on roughly the same solutions as Toyota (for instance, having the benefit of the whole system in mind, the formation of small teams, a combination of bottom-up and top-down processes, making changes in small increments). Many of the details are different from Toyota's, as expected from their separate historical origins, but they both succeed admirably as variation-and-selection processes that enable rapid adaptation to change.

That's the good news. The bad news is that decades have elapsed since the origin of the method. Despite a consulting agency (for business applications) and a nonprofit (for international aid applications), despite a book and articles in prestigious journals such as the *Harvard Business Review* and *Stanford Social Innovation Review*, despite material support from organizations such as the World

Bank, the proportion of groups that know about rapid results in either the business world or the international aid world is still small. How many decades would be required for awareness, not to speak of practice, to reach even 5 percent?

Both of my examples, Toyota and rapid results, illustrate that cultural evolution is more complex than economists like Schumpeter imagined with his phrase "creative destruction." Best practices *do* spread, but they seldom result in a "selective sweep," which is the term that biologists use when a gene rapidly evolves from mutation to fixation in a population. Instead, they spread slowly and are often opposed by other selective forces, especially disruptive forms of within-group selection when the core design principles discussed in chapter 6 are not strongly implemented. Spreading by being copied is more difficult than it seems because it is difficult to infer from the products the processes that need to be copied.

As a result of these complications, cultural adaptations end up having something similar to the geographical distribution of a biological species. They are practiced only by some groups but not by others, in some walks of life but not others, and they remain invisible outside their borders. The same is true when we examine cultural evolution at larger scales.

INNOVATION ECOSYSTEMS

So far, I have provided examples of single corporations that are adaptable by employing variation-and-selection processes. How about entire geographical regions? If we could create a map of the world that shows where technological innovations come from, it would be extremely uneven. Some regions, such as Silicon Valley, are oases of innovation, whereas most other regions are parched deserts. Silicon Valley is rare but it is not unique. Other oases exist; for example, the nation of Israel, as described by Dan Senor and Saul Singer in their book *Start-Up Nation*.[14] Like mutations and species, innovation oases have separate historical origins and spread on their own merits until they come up against boundaries, like the geographical distribution of biological species.

The absence of innovation oases in most parts of the world is not for lack of trying. On the contrary, they are the envy of the world, and countless efforts to create them have been attempted by governments, universities, and corporations. Yet these efforts almost invariably fall short of expectations. Like many products of cultural evolution, an innovation oasis works without having been designed by anyone. The active ingredients need to be discovered, even by the people who carry them out on a daily basis. They are sufficiently mysterious, at least when viewed through the lens of current theories, that they have resisted the efforts of the best and brightest who try and fail to duplicate them.

Fortunately, there is a book that begins to explain the active ingredients of innovation oases in the same way that Mike Rother explains the active ingredients of Toyota: *The Rainforest: The Secret to Building the Next Silicon Valley*, by Victor W. Hwang and Greg Horowitt.[15] Both authors have served as consultants for the creation of innovation oases (or "rainforests," as they put it) around the world, and Hwang is currently vice president of entrepreneurship at the Ewing Marion Kauffman Foundation, which has been a leader in promoting entrepreneurship for over fifty years. There is something else that distinguishes Hwang and Horowitt. In addition to their extensive professional experience, both have drunk deeply from the well of Darwinian knowledge. They employ Tinbergen's fully rounded four-question approach, which enables them to see what so many other experts are missing.

They are not blinded by the absurd conception of human nature known as *Homo economicus*, which pretends that individuals are rational actors who can be motivated entirely by money. Instead, they understand that humans are a product of genetic multilevel selection operating primarily at the scale of small groups, which has both positive and negative implications for creating a modern innovative society. On the positive side, we are innately prosocial and inclined to join in cooperative enterprises, which includes policing those who don't cooperate. On the negative side, we typically confine our prosociality to small homogeneous groups, regarding outsiders with distrust. Also, most of us are inherently risk-averse.

Gambling everything on long-shot possibilities doesn't come naturally to us.

Hwang and Horowitt are also not blinded by the pervasive assumption that an innovative culture can be explained in terms of innovative individuals. Instead, they understand that innovation is a social process requiring many different types of people cooperating with each other. The challenge is to create a society that is both highly cooperative and highly diverse. This is such an unusual combination that the ingredients fail to come together in most geographical regions, resulting in innovation deserts.

The factors that enable diverse people to cooperate in regions such as Silicon Valley and Israel are serendipitous and historically contingent. Each innovation oasis has its own story, but they typically involve situations that force people from different nationalities and walks of life to cooperate with each other. In America it was the westward expansion. In Israel it was military service. In both cases, the circle of cooperation still excludes others—Native Americans and to a large extent African Americans in the case of the American westward expansion, and Palestinians in the case of Israel—but the circle of cooperation is still sufficiently diverse to include the elements needed to create new enterprises.

In an innovation oasis, there is a high diversity of relevant skills, a high degree of social connectedness, a high degree of generosity and trust, and a high degree of dreaming about creating something that does not currently exist—which is the target of selection for an innovator. The high degree of connectedness allows bad actors to be detected and punished by more upstanding members of the community, usually without needing recourse to legal action. Someone who steals an idea or cuts an unfair deal gets a bad reputation and is excluded from future cooperative interactions. One Silicon Valley lawyer is even quoted as saying "Good businessmen don't need lawyers!" This kind of social organization is similar to hunter-gatherer tribes, where bullies are either punished or avoided and end up fending for themselves. In the case of innovation oases, however, tribal membership is open to anyone who can contribute to a current venture.

These projects might be temporary, but they are also *complex*. An apt comparison is with the movie industry—also a California specialty—where every movie requires the army of people listed in the credits to come together in a symphony of cooperation. It is a different symphony from the one that takes place in an automobile assembly plant, but a symphony all the same.

Hwang and Horowitt are practical change agents, not scientists and scholars, although they have drawn from the scholarly and scientific literature to a remarkable degree. They provide seven "Rules of the Rainforest" for those who want to create a rainforest of their own, which can readily be understood as a managed process, or policy, of cultural evolution.

RULE #1: THOU SHALT BREAK RULES AND DREAM. Achieving something new is the target of selection for groups of innovators.

RULE #2: THOU SHALT OPEN DOORS AND LISTEN. Innovation is a social process that requires cooperation.

RULE #3: TRUST AND BE TRUSTED. Trust is required to lower transaction costs. It is possible because of informal policing mechanisms.

RULE #4: EXPERIMENT AND ITERATE TOGETHER. No one is smart enough to understand a complex system and even the best theory can only narrow the field of plausible alternatives. The only way to evolve a complex system is by variation-and-selection processes.

RULE #5: SEEK FAIRNESS, NOT ADVANTAGE. Gaining advantage over other members of your group poisons collective efforts.

RULE #6: ERR, FAIL, AND PERSIST. Failure is the frontier of adaptation, for a start-up company no less than an automobile assembly plant.

RULE #7: PAY IT FORWARD. The overarching ethos of an innovation oasis is generosity, a willingness to help others without narrow expectation of gain, which is the very opposite of the conception of human nature imagined by orthodox economic theory.

THE CULTURAL EVOLUTION OF COMPLEX SYSTEMS

In this chapter I have provided three examples of adaptable societies. All of them are drawn from the business world, where between-group selection is especially intense, acting as a crucible for the cultural evolution of groups that work well in the present and adapt well in the face of environmental change.

Yet each story is more complex than the standard notion of firm selection in economics. Firms that are both adaptive in their current environments and adaptable to changing environments do arise as "cultural mutations." They do spread in competition with other firms. And they do diversify to occupy different niches in a multiple-niche economy. But decades are required and each cultural lineage comes against barriers that limit its distribution, similar to the geographical distribution of a biological species. As a result, even an example as fabled as Toyota is largely unknown outside the automotive industry, and the secrets of its success are inscrutable to many who earnestly want to copy it.

Clearly, more is needed for human groups of all sorts to adapt to change at the speed and scale that is required to solve the myriad problems of our age. The first step is to adopt the right theory. When our view of human nature is *Homo economicus* or when we think that creating an adaptive group is merely a matter of finding the most talented individuals, then we become incapable of seeing the ingredients that are actually required to create an adaptable society. We are like an engineer who is trying to build something using the wrong blueprint. It will never work, no matter how smart we are or how hard we try.

The right theory is based on the *cultural evolution of complex systems*. It notes that complex systems cannot be optimized by separately optimizing their parts. We must have in mind the performance of whole systems, which is the target of selection, and improve performance with a process of variation and selection of best practices. This is likely to work much better than laissez-faire or centralized planning. Making our societies more adaptable won't be easy, especially at larger scales, but it will be possible with the right blueprint provided here by evolutionary theory, multilevel selection, and a solid four-question approach.

Evolving the Future

I began this book with the scientist and Jesuit priest Pierre Teilhard de Chardin and his book *The Phenomenon of Man*. Teilhard regarded humanity as not just another species but also a new evolutionary process, therefore as significant in its own way as the origin of life. He imagined human evolution leading to a single global consciousness called the Omega Point that could regulate the earth as a superorganism.

In the prologue, I described this book as an updated version of *The Phenomenon of Man*. While Teilhard's book has a rich spiritual dimension, it can also be evaluated as a purely scientific thesis. Does our species represent a new evolutionary process? Can the concept of conscious evolution be justified? Do superorganisms exist? Is it theoretically possible to expand the boundary of a superorganism to include the whole earth?

These are some of the deepest questions that can be asked about evolution and the nature of humanity. Yet this book also has an immensely practical side. How do we avoid the epidemic of physical and mental illnesses that beset modern life? What are the best ways to raise our children? How can we achieve personal fulfillment? How can we cause our groups to be more effective? How can we create sustainable economies? How can we become stewards of the rest of life on earth?

For me, it is the combination of philosophical depth and practical relevance that makes an evolutionary worldview so compelling. It is easy to appreciate why Darwin was moved to write "There is grandeur in this view of life" in the final passage of *Origin of Species*. In this final chapter of my book, I will take stock of both the deep philosophical questions and the opportunities for rolling up our sleeves and employing Darwin's toolkit to make the world a better place.

THE NEED FOR EVOLUTIONARY SCIENCE

I doubt that anyone, upon serious reflection, can deny the need for scientific understanding to solve the problems of our age. Yet the attitudes of so many people about science are detached from their attitudes about evolution. A religious believer can be a science-friendly creationist. A politician can be a staunch supporter of science who doesn't dare utter the E-word. Social scientists and humanist scholars can assume that their particular disciplines are consistent with evolutionary theory, even though it was absent from their own education.

One contribution of this book, I hope, is to reveal the problem with this detachment. For all aspects of humanity, to be a scientist requires being an evolutionist. Scientists who ignore evolution run the risk of creating stockpiles of information with no interpretative framework; of asking only some of Tinbergen's four questions; or of employing interpretive frameworks that are not, in fact, consistent with evolutionary theory. Until science and evolution become more closely wedded to each other in the minds of scientists and laypeople alike, the Darwinian revolution will not be complete.

Once we consult modern evolutionary science, each of the deep philosophical questions given above can be answered in the affirmative. Taking them in order: Even though evolutionary biologists were gene-centric for most of the twentieth century, they have now gone back to basics by defining evolution in terms of variation, selection, and *heredity*, with genes as only one of several mechanisms of heredity. Cultural transmission represents another mechanism of heredity in many species, one that has become vastly more powerful in humans with the advent of symbolic thought. Teilhard was on the mark when he wrote that our species represents a new (or at least greatly expanded) evolutionary process.

Turning to the question of conscious evolution, it was dogma during most of the twentieth century that evolution is an undirected process lacking any kind of purpose. Specifically, the dogma held that variation is random with respect to the traits that are selected by the environment. Giraffes stretching to reach high foliage do not mysteriously cause their offspring to become taller. Their offspring are both taller and shorter, with differential survival and reproduction explaining why the giraffe population evolves to become taller over time.

Historically, it is easy to understand why this dogma became so strong, to reject fuzzy ideas of progressive evolution associated with Herbert Spencer and others. Also, at first it was difficult to see how the experiences of an organism could influence the genes in the organism's reproductive cells. Today we know that this is possible after all—not by changing the presence of the genes in the reproductive cells, but by changing patterns of gene expression (epigenetics). In other words, the experiences of your parents resulted in some of their genes being up-regulated and others being down-regulated. To a degree, these patterns of gene expression were transmitted to you. The way you are can even bear the traces of the experiences of your grandparents and great-grandparents.

When we turn to individual learning and cultural transmission as evolutionary processes, there is obviously a strong directed component. Thus, the dogma that evolution is undirected is largely an artifact of gene-centric evolutionary biology. Also, learning and cultural transmission can radically alter the course of genetic evolution. Some human populations evolved the genetic ability to digest lactose as adults only because they had previously learned to domesticate livestock. In this fashion, the slow process of genetic evolution follows where the fast process of cultural evolution leads.

The bottom line for our purposes is that there is nothing wrong with the concept of conscious evolution. Consider the process of conscious decision-making. There is a clear objective for evaluating alternative options, which is the target of selection in evolutionary terms. The variation part of the evolutionary process includes both a directed and an undirected component. We don't suggest options at random; typically, we are guided by one set of expectations or another. On the other hand, some options do appear to "come out of nowhere," and these are often the ones that are chosen. That's what brainstorming is all about. One way to demonstrate the importance of the random component is by giving the same problem to a number of decision-making groups. They usually come up with different solutions, just as the lines of *E. coli* in Richard Lenski's experiment evolved different ways to digest glucose.

Evolutionary algorithms in computer science illustrate the same point. Some problems, such as how a traveling salesman should minimize the length of his path through different cities, are notoriously difficult to solve because there are so many combinatorial possibilities. One way to proceed is to represent different options (each path through the cities) as a string of information, like genes on a chromosome, and to select them on the basis of path length. Then variation is created by mutating the strings and recombining them with each other, emulating the process of genetic recombination. Numerous "generations" of this process do a good job of finding the shortest paths. The whole process is consciously designed to solve a specified problem, but it still counts as an evolutionary process.

Once we become comfortable with the concept of conscious evolution, then the need to design our personal and cultural evolutionary processes becomes clear, similar to designing an evolutionary algorithm on a computer. This is what biologists call "the evolution of evolvability."[1] If we don't become wise managers of evolutionary processes, then evolution will still take place but will lead to outcomes that are not aligned with our normative goals. Selecting economic practices to maximize gross national product (GNP) is one of many examples that could be cited. Even if we manage to succeed, it turns out that GNP is not a good proxy for societal welfare. We chose the wrong target for selection and evolution became the problem rather than the solution. Examples like this can be listed almost without end.

Do superorganisms exist? The answer to this question is unequivocally "yes." An entity counts as an organism to the degree that it is a unit of selection in multilevel evolution. Every entity that is currently described as an organism, such as a bacterial cell, a nucleated cell, or a multicellular organism, is a highly regulated society of lower-level units that behave as they do only because they were selected as groups. To the degree that selection also operates within groups, these same lower-level units can evolve to be cancerous. Eusocial insect colonies qualify as organisms because they are the primary units of selection, even when the colony members are physically dispersed. The fact that these ideas apply with equal force to human genetic and cultural evolution is a scientific breakthrough of the first rank.

Can the boundary of the human superorganism be expanded, even to embrace the entire earth? It already has expanded to a considerable degree over the last ten thousand years, resulting in the mega-societies of today. Every nation can improve its governance, but the degree of cooperation and coordination that takes place would be inconceivable to our hunter-gatherer ancestors. If we want the whole earth to become a superorganism, then multilevel selection theory tells us exactly what to do: make planetary welfare the target of selection. This is easier said than done, but clearly establishing the target of selection is the first step, especially given competing narratives that portray the unregulated pursuit of lower-level interests as the way to benefit the common good.

In light of recent developments in modern evolutionary science, we can now answer in the affirmative every deep philosophical question posed by Teilhard in *The Phenomenon of Man*. This book could not have been written twenty years ago. One reason that the Darwinian revolution is not yet complete is because evolutionary biologists became so restricted in their view of evolution, ceding the study of non-genetic evolutionary processes to other disciplines. That is now rapidly changing, but there is a lot of catching up to do.

FROM A THEORY TO A WORLDVIEW

In this book, I have used the phrase "evolutionary worldview" more often than "evolutionary theory." The difference is that a theory can only tell us what *is*, while a worldview can tell us *how to act*. Teilhard described the Omega Point not only as a scientific possibility, but also as one worth working toward. This message was so inspiring that his book continues to be widely read, even after it was forgotten by most scientists.

We are usually so focused on our disagreements that we lose sight of our common ground. Nearly all of us think that it's right to avoid disruptive self-serving behaviors and to cooperate with others to achieve common goals, at least within a defined moral circle. That is why I could invite you to describe a morally perfect individual in chapter 4, knowing in advance what you were going to say. We disagree on how we define our moral circles, on what needs to be done to achieve common goals, and also on the narratives that convey our moral ideals, but these are relatively superficial compared to our shared moral psychology, which is designed to help us work together in groups.

Not only do we have a shared moral psychology, but it is relatively easy to agree upon the whole earth as the ultimate moral circle, as strange as this might seem. The great novelist Joseph Conrad once said that he enjoyed writing books about the sea because life on a ship is so morally simple. It's obvious to everyone on the ship that the common goal is to remain afloat. Space fantasies such as *Star Trek* and *Star Wars* have the same appeal. Get anyone to imagine the whole earth as like a single ship, and that person will start regarding the whole earth as the appropriate moral circle.

Religions are sometimes described as eternally at war with each other, but this is far from true. Religious wars, like all wars, are highly contextual. As cultural superorganisms, religious groups are capable of employing both aggressive and cooperative social strategies. Which strategy prevails depends upon the socio-ecological conditions. If you want to stop wars, think like an evolutionary ecologist by providing the right conditions for peaceful social strategies to win the Darwinian contest.

Moreover, the contemplatives of all major religious traditions—from Christian monks in their monasteries to Buddhist hermits in their caves—converge upon a common awareness of rich interconnectedness. Once life is seen as a vast interconnected system, certain ethical conclusions follow. Specifically, the futility of one part of the system attacking another part of the system is revealed. Systemic problems require systemic solutions. Pope Francis's encyclical on the environment, *On Care for Our Common Home*, reaches out to all people. His Holiness the Dalai Lama even wrote a book titled *Beyond Religion: Ethics for a Whole World*.

Environmental thinkers reach the same conclusion. They achieve their appreciation of rich interconnectedness through the study of nature, but the same ethical conclusions follow. This is why environmental philosophers such as Arne Naess, who coined the term "deep ecology," have a spiritual quality without invoking any gods.

Even economists and politicians can achieve common ground on a whole earth ethic, as strange as this might seem. Those who preach "greed is good" aren't saying that it's OK to be greedy even if the world falls apart as a result. They're saying that greed is good for society, including the global economy. The current U.S. president, Donald Trump, isn't saying that he wants to put America first at the expense of other nations or the world. At least when he's justifying his policies on the world stage, he's saying that if every nation attempts to put itself first, then they will all strike the right kind of deals for the common good. He might be wrong about his strategy, but rhetorically he is not abandoning a whole earth ethic. Anyone who truly abandoned a whole earth ethic on the world stage would be properly regarded as just plain immoral.

Against this background, an evolutionary worldview makes two contributions toward actually achieving a whole earth ethic, as opposed to just wanting it when in the right frame of mind. First, it provides strong scientific support for the need to make the welfare of the whole earth the target of selection in the formulation of policies, which in turn can orchestrate the formulation of policies for all lower-level units. At the same time, it removes scientific support for the main alternative narrative, which is that the unregulated pursuit of lower-level self-interest robustly benefits the common good. To put it bluntly, when it comes to policy formulation, laissez-faire is dead.

The second contribution is to provide an alternative to centralized planning, which is also dead as a way of designing complex systems. Policy formulation must be a conscious evolutionary process that first chooses the right target of selection, then monitors planned and unplanned variation, and then replicates best practices, realizing that implementation might be highly contextual. Engineers have already reached this conclusion for the complex systems that they attempt to design.[2] How could the whole earth and its many subsystems be any different?

Another insight from an evolutionary worldview is that change must be incremental. It must be possible to get there from here. Optimally, progress can be like climbing a mighty adaptive peak, where every step takes us upward, rather than needing to descend into adaptive valleys along the way.

WHAT WE CAN DO

We—the writer and readers of this book—are a diverse group. Some of us are in a position to make things happen at a large scale. All of us are in a position to make things happen at *some* scale, if only in our own lives and our immediate surroundings. Fortunately, acting upon a whole earth ethic is not a matter of sacrificing your welfare for the benefit of the planet. It can be done in a way that helps you to thrive as an individual by taking part in groups that are richly rewarding and powerful agents of change at a larger scale.

A remarkable example, with lessons for all of us, is an ecovillage called Dancing Rabbit located in rural Missouri.[3] Dancing Rabbit does not strive to set itself apart from the rest of society. Instead, it models itself after a New England village, but one that consciously strives to provide a good life while using a tenth of the resources of the average American. Here is its mission statement.

> To create a society, the size of a small town or village, made up of individuals and communities of various sizes and social structures, which allows and encourages its members to live sustainably.*
>
> To encourage this sustainable society to grow to have the size and recognition necessary to have an influence on the global community by example, education, and research.
>
> *Sustainably: In such a manner that, within the defined area, no resources are consumed faster than their natural replenishment, and the enclosed system can continue indefinitely without degradation of its internal resource base or the standard of living of the people and the rest of the ecosystem within it, and without contributing to the non-sustainability of ecosystems outside.

The Rabbits, as they call themselves, take their mission seriously. Newcomers are required to sign a covenant that regulates vehicle use, fossil fuel use, agricultural practices, building materials, electrical and water use, and waste disposal. They also commit to nonviolent conflict resolution and to contribute a portion of their time and income to the community. If they repeatedly break the covenant, the village has the right to exclude them.

These norms regulating sustainable living are as strict as can be. Yet, for all other aspects of their lives, the Rabbits are as tolerant as can be. Any religion (including none). Any sexual orientation. Any ethnicity. Any living arrangement. Also, the mission statement and covenant do not prescribe exactly how to abide by the sustainability norms. Instead, they encourage experimentation. They have designed the village as an evolutionary process with a clear target of selection that has the welfare of the whole earth in mind.

Dancing Rabbit illustrates one of the main take-home messages of this book: the importance of small nurturing groups for both individual well-being and efficacious action at a larger scale. I have visited the village and can attest to its high quality of life. The first thing I noticed was the absence of cars. The communally owned cars are parked on the edge of the village and the buildings are separated by walking paths. The second thing I noticed was that children were running around playing on their own without needing to be supervised. Who needs to be a helicopter mom or dad when there is no danger of being hit by a car or abducted by strangers? The third thing I noticed was the abundance of physical activities. People were doing things with their bodies, mostly outdoors, such as tending their gardens, building their houses, maintaining the common grounds. These things used to come for free in olden times but must be constructed in modern life. It's nearly impossible to construct them as an individual but they can be constructed as a group.

Life at Dancing Rabbit is not easy. In many ways, the Rabbits suffer the hardships of pioneer life, but these can also be immensely rewarding. In their own internal assessment, 88 percent of the Rabbits responded either "good" or "extremely good" to the question "Do you think that Dancing Rabbit is a good place to live?" Many comment on the village as a great place for personal growth. Strict norms aren't resented when they represent your own values and the norm for tolerance in all other respects provides room for experimentation.

So, as a strong nurturing group, Dancing Rabbit contributes to individual well-being. The village also contributes to efficacious action at a larger scale. The Rabbits are research-oriented and conduct a careful audit of their ecological footprint. Compared to national averages, they create 7 percent of municipal solid waste; less than 10 percent of car use; 5.5 percent of propane gas consumption; 17.7 percent of electrical use, generated almost entirely on site; 7.5 percent of water use. All on an average salary of about $10,000 per year, which gets you abject poverty anywhere else in America. Finally, not only do they provide a model for other groups, but they actively promote their model through their various educational and outreach activities. The group is far, far more powerful than any member could be on their own.

With two Norwegian colleagues, Bjørn Grinde and Ragnhild

Bang Nes, and one of my graduate students, Ian MacDonald, I have conducted a survey of over a hundred intentional communities that belong to a consortium called the Fellowship of Intentional Communities (FIC).[4] Some were rural, others urban. Some were religious, others secular. Some were ecovillages, others had different focuses, such as eldercare. As a sample, all of them illustrated the concept of small groups leading to both individual well-being and efficacious action at a larger scale. They scored near the top on measures of satisfaction and meaning of life that have been given to many sample populations around the world. They reported "community" as their reason for joining, their current satisfaction, and what they wanted more of by way of future change.

In addition to their high average values, the intentional communities in our sample also *varied* in how well they achieved their objectives. As we predicted, this variation was based in part on how well they had implemented the core design principles. The average intentional community performs well, but can perform even better with a little evolutionary know-how.

Can't drop everything to join an intentional community? You don't need to. You already participate in numerous group activities, and each one can benefit from the same principles illustrated by Dancing Rabbit. Here is a guide that all of us can follow.

1. BECOME MORE MINDFUL OF YOUR OWN VALUES AND GOALS. Changing the world begins with adopting a whole earth ethic for ourselves and translating it into what we can do locally. I highly recommend that you learn more about science-based mindfulness techniques such as ACT (described in chapter 7). You need to become more conscious about your own evolution to bring it into alignment with your normative goals! If you wish to consult with a therapist or a coach, visit the website of the Association for Contextual Behavioral Science.[5]

2. BECOME A MEMBER OF WORTHWHILE GROUPS. We are designed by evolution to be members of small coop-

erative groups where we can be known and respected for our actions. We are also designed to participate in numerous such groups simultaneously, each oriented around a different set of activities. The societies of our hunter-gatherer ancestors were not so different from the way that we split our time between our various group activities. As much as you can, choose to participate in groups that reflect your values and goals.

3. MAKE YOUR GROUPS STRONG BY IMPLEMENTING THE CORE AND APPROPRIATE AUXILIARY DESIGN PRINCIPLES. Even though we are designed by evolution to live in small groups, that doesn't mean that we instinctively implement the right design principles. Instead, groups *vary* in their implementation, and even when they do a good job, they are not necessarily consciously aware of what they are doing right. This leaves tremendous room for almost all groups to improve by becoming more mindful about *their* evolution, just as you need to become more mindful about *your* evolution.

4. MAKE YOUR GROUP A HEALTHY CELL IN A MULTICEL-LULAR SOCIETY. Becoming a participant in small and well-organized groups is likely to do wonders for your personal welfare, but those groups must also contribute positively to society at a larger scale. The fact that the core design principles are scale-independent provides a blueprint for how to proceed. As a cooperative superorganism, your group needs to seek out other cooperative superorganisms, avoid the hazards of more predatory superorganisms, and work to establish the core design principles at larger scales. You do not require the permission of any authorities to start doing this. The B Corp movement described in chapter 6 provides an example of enlightened businesses doing it on their own. However, you should be on the lookout for authorities (such as governments) that are also enlightened enough

to provide the right kind of top-down assistance, as the city of Buffalo did for its block clubs as described in chapter 6.

How about if you are lucky enough to make a difference at a larger scale? What if you are a politician, policy expert, CEO of a business, or board member of a philanthropic foundation? Then you are in a position to provide the top-down assistance to which I just alluded. Avoid the temptation of centralized planning. You and your expert consultants are not smart enough to know how to improve the complex system that is under your care. You should implement the core design principles, which are fundamentally inclusive. You should also implement evolutionary processes that identify the right targets of selection, monitor planned and unplanned variation, and replicate best practices, knowing that their implementation will be highly sensitive to context. Everyone within the complex system under your care will need to take part, as in the examples of Toyota, the rapid results method, and innovation ecosystems that I recounted in chapter 9. You will have far more success, recognition, and personal fulfillment than you would pretending to be a Master of the Universe.

Here is something else that everyone can do: learn more about the evolutionary worldview. Think of this book as a portal to a world that invites much more exploration. The number of people who have already gone through the portal is large, although still a tiny fraction of the worldwide population. One of my greatest delights as a teacher and writer is to escort people through the portal. They repeatedly tell me how an evolutionary worldview has gotten "under their skin," how they talk about it to their friends and apply it to all aspects of their lives. They are clearly seeing the world in a new way with no more than a basic introduction, which of course can be deepened with further study. This is not just my own experience but also that of others who teach and write about evolution in the same way. It is the theory presented in an appropriately general form, not the teacher or writer, that accomplishes the transformation.[6]

Speaking for myself, I have come a long way from my start as a graduate student in 1973, the year that Dobzhansky wrote "Nothing in biology makes sense except in the light of evolution." I'm proud to have played a role in expanding the concept of "biology" to include everything associated with the words "human" and "culture." I am among the lucky ones in a position to make a difference at a larger scale, thanks to the Evolution Institute,[7] which is currently the only think tank that explicitly draws upon evolutionary theory in its formulation of public policy. Two EI projects that are especially amenable to participation are Prosocial[8] and the TVOL1000.[9] Prosocial is a practical method for helping groups evolve their futures and take part in the construction of cooperative multigroup ecosystems, based on the ideas presented in this book. The TVOL1000 is a group that supports the EI's online magazine *This View of Life*[10] and forms into action groups on topics such as business, education, parenting, and health.

I have learned from my own efforts that no matter what topic is chosen for policy formulation, a rich array of best practices already exists—such as the Good Behavior Game for education, Buffalo's West Side Neighborhood Association, evangelical cell ministries, and the rapid results method for businesses. Each of these best practices arose as a "cultural mutation," often by happenstance, and spread on the basis of its success. That's what cultural evolution is all about. However, their spread was often slow, requiring decades, and frequently came to a stop against various boundaries, similar to the geographical distributions of biological species. Also, many of the best practices work without a clear understanding of why they work. The explanations that exist are often expressed in terms of narrow domains of knowledge that are difficult to transfer to other domains.

As a result, while we should always be on the lookout for best practices and learn from their success, more is needed to solve the problems of our age as quickly as we need to and at a planetary scale. The main contribution of evolutionary theory is not to discover solutions that have never been tried before, but to provide a general explanatory framework that identifies why best practices work and how they can be spread across all domains of knowledge and policy applications. That was the synthesis that organized the biological sciences in my youth and is now in progress for everything associated with the words "human" and "culture."

There is no question in my mind that "this view of life" is the wave of the future. The main question is how soon it will arrive and whether it will be in time to avert the potential disasters that confront us. I look forward to the day when the whole world will be saying, along with Thomas Huxley, how stupid of us not to have thought of that.

Notes

1. See Amir (2007) and King (2015) for excellent biographies of Teilhard. I devote a chapter to him in my book *The Neighborhood Project* (Wilson 2011) and also was among several interviewed about him by Krista Tippett on her radio show *On Being:* www.onbeing .org/programs/ursula-king-andrew-revkin-and-david-sloan -wilson-teilhard-de-chardins-planetary-mind-and-our-spiritual -evolution/.

INTRODUCTION: THIS VIEW OF LIFE

1. It is remarkable that such a powerful theory can be described in a single paragraph. I am assuming that my reader knows at least this much about Darwin's theory. For a longer introduction, please consult my earlier book *Evolution for Everyone: How Darwin's Theory Can Change the Way We Think About Our Lives* (Wilson 2007).
2. Dobzhansky (1973).
3. Wilson (1973).
4. I explore this theme in an interview with Peter and Rosemary Grant, the couple who is world-famous for studying the Galápagos finches, in the online magazine *This View of Life* (hereafter TVOL): https://evolution-institute.org/article/when-evolutionists-acquire -superhuman-powers-a-conversation-with-peter-and-rosemary -grant/.
5. The history of sociology and cultural anthropology in relation to evolutionary theory are the topics of my TVOL interviews with

Russell Schutt (https://evolution-institute.org/article/why-did
-sociology-declare-independence-from-biology-and-can-they
-be-reunited-an-interview-with-russell-schutt/) and Robert Paul
(https://evolution-institute.org/article/cultural-anthropology
-and-cultural-evolution-tear-down-this-wall-a-conversation-with
-robert-paul/).

6. Beinhocker (2006).
7. Veblen (1898).
8. The seminal work is Nelson and Winter's (1982) "An Evolutionary
   Theory of Economic Change."
9. Frank, R. (2011).

1. DISPELLING THE MYTH OF SOCIAL DARWINISM

1. https://www.youtube.com/watch?v=s56Z5lofYV0.
2. Hodgson (2004). I highly recommend Hodgson's other work listed
   in the bibliography as an example of an evolutionary worldview
   leading to a positive agenda for social change.
3. See Bashford and Chaplin (2016) for a good modern biography of
   Malthus and his work.
4. See Francis (2007) for a good modern biography of Spencer and
   his work.
5. Of the many biographies of Darwin, I recommend the two volumes of Janet Browne (1995, 2002).
6. Darwin (1871), vol. 1, p. 162.
7. See Brookes (2004) for a good modern biography of Galton and his
   work.
8. http://galton.org/essays/1870–1879/galton-1872-fort-rev-prayer
   .pdf.
9. See Barr (1997) for a centennial assessment of Huxley, whose relationship with Darwin is well told by Browne (1995, 2002).
10. See Provine (2001) for a fascinating account of this period of evo-
    lutionary thought.
11. See Paradis and Williams (2016) for modern commentaries in
    addition to Huxley's original essay. "I doubt whether": p. 23.
12. Ibid., p. 29.
13. Ibid., p. 95.

14. See Dugatkin (2011) for a good modern biography of Kropotkin and his work.
15. A. Novoa (2016) describes how Darwin's theory was refracted through the lens of different cultures.
16. Richards (2013); the title essay is excerpted in Wilson and Johnson (2016b).
17. All quotes in this section are taken from Richards (2013).
18. For an in-depth discussion of Dewey, see my interview with the philosopher Trevor Pearce in Wilson and Johnson (2016b).
19. Dewey, J. (1910), pp. 1–2.
20. See Jablonka and Lamb (2006), Richerson and Boyd (2005), and Henrich (2015) for more on these topics.
21. See my interview with the sociologist Russell Schutt in Wilson and Johnson (2016b) and books by Geoffrey Hodgson in the Literature Cited section for more on the history of sociology and other disciplines in the social sciences in relation to evolution.

2. DARWIN'S TOOLKIT

1. I first explored this theme in a 2009 blog: http://scienceblogs.com/evolution/2009/10/20/goodbye-huffpost-hello-science/.
2. Tinbergen's classic article outlining the four questions was "On Aims and Methods of Ethology" (Tinbergen 1963). His book *The Curious Naturalist* (Tinbergen 1969) is wonderful to read and shows how he employed his "four-question" approach in his own work. Bateson and Laland (2013) provides a modern appreciation and update.
3. The *Wikipedia* entry for laissez-faire documents the origin of the phrase: https://en.wikipedia.org/wiki/Laissez-faire#Etymology_and_usage.
4. See Bodkin (1990) and Kricher (2009) for reviews of the balance of nature concept.
5. For more on the need to go beyond laissez-faire in economics, see Wilson and Kirman (2016), Colander and Kupers (2014), and Wilson and Gowdy (2014).
6. A classic article by Stephen Jay Gould and Richard Lewontin (1979) argues against interpreting everything as an adaptation.

My TVOL interview with Richard Lewontin provides fascinating historical context: https://evolution-institute.org/article/the -spandrels-of-san-marco-revisited-an-interview-with-richard-c -lewontin/.

7. Lloyd, Wilson, and Sober (2014) provides a tutorial on evolutionary mismatch.

8. It is a striking fact about Tinbergen's function question that it provides a solid scientific foundation for holistic statements such as "the parts permit, but do not cause, the properties of the whole." See Wilson (1988) and Campbell (1990) for more.

9. See my *This View of Life* interview with Lenski for more: https:// evolution-institute.org/article/evolutionary-biologys-master -craftsman-an-interview-with-richard-lenski/.

3. POLICY AS A BRANCH OF BIOLOGY

1. See Schwab (2012) for a good trade book on the evolution of eyes.

2. Gómez et al. (2009).

3. My account of eye development is based largely on the website *Eye, Brain, and Vision*, which is maintained by David Hubel, one of the giants of the field: http://hubel.med.harvard.edu/index.html.

4. Morgan et al. (1975).

5. This and the rest of the examples in this section are drawn from Sherwin et al. (2012), Goldschmidt and Jacobsen (2014), and Cordain et al. (2002).

6. The term "natural experiment" refers to a comparison that can be made while holding other factors constant, similar to a planned experiment. In this case, comparing ethnic Chinese living in Singapore and Australia allows the amount of time spent outdoors to be evaluated while holding other behaviors associated with being Chinese constant. Of course, a natural experiment is seldom as well controlled as an actual experiment.

7. Nuland (2003).

8. Gaynes and the American Society for Microbiology (2011).

9. For a good popular account of microbiomes, see Yong (2006). The academic literature that I consulted for this section includes

Bloomfield et al. (2016), Dethlefsen et al. (2007), Flandroy et al. (2018), Hanski et al. (2012), Miller and Raison (2016), Rook (2012, 2013), Rook and Lowry (2008), and Sender et al. (2016).

10. Sender et al. (2016).

11. Miller and Raison (2016).

12. Visit Graham Rook's website for this and other interviews, along with many other resources to learn about the hygiene hypothesis: https://www.grahamrook.net.

13. Hanski et al. (2012); see also Rook (2013) and Flandroy et al. (2018).

14. This phrase was coined by John Bowlby, a pioneer in the study of human development from an evolutionary perspective. See Bowlby (1990) for a review of his work.

15. Lickliter and Virkar (1989).

16. For an excellent trade book on this material, see Gray (2013). For an academic review of the material in this section, see Bjorklund and Ellis (2014). For applications to early childhood education, see Gray (2013) and Geary and Berch (2016). For an accessible online article, see https://www.psychologytoday.com/us/blog/freedom -learn/201505/early-academic-training-produces-long-term -harm.

17. Darling-Hammond and Snyder (1992).

18. Schweinhart and Weikart (1997).

19. Barr (2013).

20. DeLoache et al. (2010).

21. Zimmerman et al. (2010).

22. Radesky et al. (2014).

23. Gray (2013).

4. THE PROBLEM OF GOODNESS

1. For a concise book-length account of this material, please consult my book *Does Altruism Exist? Culture, Genes, and the Welfare of Others* (Wilson 2015).

2. Wilson and Wilson (2007).

3. In addition to the epigenetics that takes place during the lifetime of a single organism, which is required for cell differentiation, epi-

genetics can also be transgenerational and therefore an inheritance system in its own right that operates alongside genetic inheritance. See Jablonka and Lamb (2006) for more on that topic.

4. Martincorena et al. (2015).

5. See my *This View of Life* interview with cancer researcher Athena Aktipis for more on the novelty of evolutionary theory for cancer research: https://evolution-institute.org/article/the-evolutionary -ecology-of-cancer-an-interview-with-athena-aktipis/.

6. Muir (1995), Craig and Muir (1995), Muir et al. (2010).

7. Margulis (1970).

8. Maynard Smith and Szathmáry (1995, 1999).

9. Szathmáry and Maynard Smith (1997), Higgs and Lehman (2015).

10. Holldobler and Wilson (2008), Seeley (1996, 2010).

11. Seeley (2010).

12. Ehrenpreis et al. (1978).

13. Boehm (1999, 2011).

14. https://evolution-institute.org/article/evolution-and-morality-1 -simon-blackburn/.

5. EVOLUTION IN WARP DRIVE

1. For a highly readable account of how the immune system works, see Sompayrac (2008).

2. Benard (2008).

3. For a biography of Skinner, see Bjork (1993). For a history of behavioral therapies as told by their founders' personal histories, see O'Donohue et al. (2001).

4. For histories of the cognitive revolution, see Miller (2003) and Gardner (1985).

5. The most influential book on evolutionary psychology is Barkow et al. (1992). For a modern assessment of the field written for a general audience, see the special edition of TVOL titled *What's Wrong (and Right) About Evolutionary Psychology*: https://evolution -institute.org/wp-content/uploads/2016/03/20160307_evopsych_ ebook.pdf.

6. T. D. Wilson (2015), Hayes and Smith (2005).

7. See Ellis et al. (2017) for a review of adaptations to stressful envi-

ronments in both humans and nonhumans. Ellis et al. (2012) applies this concept to risky adolescent behavior in humans.

8. For this and subsequent examples in this section, see Ellis et al. (2017) for a review.

9. Key books accessible to a general audience include Jablonka and Lamb (2006), Henrich (2015), Laland (2017), and Richerson and Boyd (2005).

10. Turchin (2015).

11. For a history of cultural anthropology see my TVOL interview with Robert A. Paul: https://evolution-institute.org/article/ cultural-anthropology-and-cultural-evolution-tear-down-this -wall-a-conversation-with-robert-paul/.

12. I develop this melting pot theme in a TVOL essay, "The One Culture: Four New Books Indicate That the Barrier Between Science and the Humanities Is at Last Breaking Down": https://evolution -institute.org/focus-article/the-one-culture/.

13. Wilson (2005a).

14. Chudek et al. (2012).

15. Gregory and Webster (1996).

6. WHAT ALL GROUPS NEED

1. I tell this story in my book *The Neighborhood Project: Using Evolution to Improve My City, One Block at a Time* (Wilson 2011). Please visit the Evolution Institute website to learn about our many important projects: https://evolution-institute.org.

2. Campbell (1994).

3. See Wilson and Kirman (2016) and Wilson and Gowdy (2014) for discussions of evolutionary theory as a new paradigm for economics. The online magazine *Evonomics.com* features much of this material for a general audience.

4. Wilson (1988).

5. Swaab et al. (2014).

6. Boehm (2011).

7. http://freakonomics.com/2009/10/12/what-this-years-nobel-prize -in-economics-says-about-the-nobel-prize-in-economics/.

8. Hardin (1968).

9. Ostrom's best-known book is *Governing the Commons: The Evolution of Institutions for Collective Action* (Ostrom 1990). Her more recent contributions include Ostrom (2010a, 2010b, 2013).

10. Cox et al. (2010).

11. I tell this story in "Evonomics," a chapter of *The Neighborhood Project* (Wilson 2011).

12. Wilson, Ostrom, and Cox (2013).

13. Csikszentmihalyi et al. (1993).

14. Wilson, Kauffman, and Purdy (2011).

15. Embry (2002) provides an excellent discussion of the Good Behavior Game and how it works. See also Wilson et al. (2014), Domitrovich et al. (2016), and references cited below.

16. Barrish et al. (1969).

17. Kellam et al. (2011, 2014).

18. Domitrovich et al. (2016). Visit https://paxis.org for recent updates and to see how the GBG can be implemented in schools in your region.

19. Sampson (2003).

20. Oakerson and Clifton (2011); discussed in Wilson, Ostrom, and Cox (2013).

21. Jacobs (1961), Putnam (2000).

22. My book *Darwin's Cathedral: Evolution, Religion, and the Nature of Society* (Wilson 2002) helped to initiate the study of religion from a modern evolutionary perspective. For a recent progress report, see Wilson et al. (2017) and Sosis et al. (2017).

23. I devote a full chapter of my book *Darwin's Cathedral* (Wilson 2002) to Calvinism.

24. Cho (1981).

25. Jones (2012).

26. https://www.youtube.com/watch?v=3pS2LthVoac.

27. This section is based on an Evolution Institute report, *Doing Well by Doing Good* (Wilson et al. 2015).

28. Pfeffer (1988).

29. Welbourne and Andrews (1996).

30. Jackall (2009).

31. See Haidt's website: http://ethicalsystems.org.

32. Grant (2013).

33. Nuttall (2012).

34. https://www.bcorporation.net.

35. Chen and Kelly (2014).

7. FROM GROUPS TO INDIVIDUALS

1. Visit Coan's Affective Neuroscience Laboratory website for an overview of his research and links to his academic publications: https://affectiveneuroscience.org.

2. Coan et al. (2006), Coan and Sbarra (2015), Beckes and Coan (2011), Beckes et al. (2013).

3. The most recent edition of *Introduction to Behavioural Ecology* is Davies, Krebs, and West (2012).

4. http://circleofwillispodcast.com/bonus-david-sloan-wilson-interviews-me.

5. Bhalla and Proffitt (1999).

6. See Gallace and Spence (2010) for an overview.

7. One of Skinner's most influential articles had this title (Skinner 1981) and was written a few years before his death.

8. Trivers (1972), Burt and Trivers (2006), Daly and Wilson (1988).

9. Haig (1993, 2015).

10. The importance of alloparenting is emphasized by Sarah Hrdy (2011) in her book *Mothers and Others: The Evolutionary Origins of Human Understanding.*

11. Ofek (2001).

12. Turchin (2015).

13. See Biglan (2015) for more on this topic.

14. Triple P website: http://www.triplep.net/glo-en/home/.

15. Prinz et al. (2009).

16. Hayes and Smith (2005), Hayes et al. (2011).

17. Grunbaum (1984). For an evaluation of modern long-term psychoanalytic methods, see Smit et al. (2012).

18. Muto et al. (2011).

19. Jeffcoat and Hayes (2012).

20. Wilson and Hayes (2018), Hayes et al. (2017), Wilson et al. (2014).

21. This scenario does not deny the fact that some forms of psycho-pathology are caused by organic dysfunctions.

22. Provine (1986) provides an interesting discussion of the adaptive landscapes in his biography of its originator, Sewall Wright. For a more recent treatment, see Svensson and Calsbeek (2012).

23. The concept of symbolic thought as a mechanism of inheritance is explored by Jablonka and Lamb (2006) and Wilson et al. (2014).

24. Pennebaker and Seagal (1999), Pennebaker (2010).

25. Walton and Cohen (2011).

26. The concept of a symbotype is significantly different from the con-cept of memes popularized by Richard Dawkins in his 1976 book *The Selfish Gene*. Dawkins imagined memes as closely analogous to his conception of genes, which itself has turned out to be prob-lematic. The concept of a symbotype is functionally analogous to a genotype and shares important properties such as combinatorial diversity (similar to genetic recombination), but that's as far as the comparison goes. See Henrich et al. (2008) for an informative dis-cussion of how the modern study of cultural evolution differs from the meme concept.

27. Polk et al. (2014).

28. If you would like to learn more about ACT and possibly work with an ACT trainer, visit the website of the Association for Contextual Behavioral Science: https://contextualscience.org/acbs.

29. See chapter 16 of Wilson and Hayes (2018).

8. FROM GROUPS TO MULTICELLULAR SOCIETY

1. Ellis (2015).

2. Turchin (2015).

3. For a highly readable history of economics oriented toward TVOL, see Beinhocker (2006).

4. Mandeville ([1714] 1988).

5. Although Adam Smith is credited with inventing the metaphor of the invisible hand, he used it only three times in all of his writing and it does not represent the full corpus of his thought. See Wright (2005) for an informative critique.

6. Wilson and Gowdy (2014).

7. For a sample of academic articles on shame from a genetic and cultural evolutionary perspective, see Fessler (2004), Nichols (2015), Folger et al. (2014), Tanaka et al. (2015), and Sznycer et al. (2016).

8. Smail (2008).

9. Visit Peter Turchin's website for more: http://peterturchin.com.

10. Turchin (2010, 2015, 2016).

11. While Dawkins (2006) interprets most religious beliefs as parasitic memes, most enduring religious beliefs and practices are better understood as for the good of the religious group (Wilson 2002). However, parasitic memes are still a theoretical possibility, and I interpret one short-lived religious movement, Millerism, which was a precursor to Seventh-day Adventism, as parasitic in Wilson (2011), chapter 18.

12. The Evolution Institute is making a special study of Norway as a case study of cultural evolution leading to a high quality of life: https://evolution-institute.org/projects/norway.

13. Acemoglu and Robinson (2012).

14. Pickett and Wilkinson (2009).

15. Fukuyama (2012).

16. O'Brien (2018).

17. This theme is developed in my *This View of Life* conversation with the systems engineer Guru Madhavan: https://evolution-institute.org/systems-engineering-as-cultural-group-selection-a-conversation-with-guru-madhavan/.

18. My TVOL essay with the Norwegian biologist Dag Hessen, "Blueprint for the Global Village," explores this theme: https://evolution-institute.org/focus-article/blueprint-for-the-global-village/.

9. ADAPTING TO CHANGE

1. See Wilson and Kirman (2016) and Colander and Kupers (2014) on the special challenges of evolving complex systems.

2. Colander and Kupers (2014).

3. The concept of science as an evolutionary process has a long history in philosophical thought. In addition, science can be studied as

a product of cultural evolution. For example, see Campbell (1974), Hull (1990), Shapin (1995), and McCauley (2011).

4. Wagner (1996).
5. Vasas et al. (2011).
6. Yasha Hartberg and I have explored the idea that written sacred texts were an important advance in cultural evolution and can be studied using the same techniques as biological inheritance systems: Hartberg and Wilson (2016).
7. Rother (2009).
8. For an excellent basic tutorial on complexity theory, see Gleick (1987).
9. My account of Toyota is based largely on *Toyota Kata* by the systems analyst Mike Rother (2009).
10. See Rother (2017) and Rother et al. (2017) for extensions beyond the domain of business.
11. Schaffer and Ashkenas (2007).
12. Hedges (2002), Solnit (2009).
13. Matta and Ashkenas (2003), Matta and Morgan (2011).
14. Senor and Singer (2009).
15. Hwang and Horowitt (2012).

10. EVOLVING THE FUTURE

1. Wagner (1996).
2. See my *This View of Life* conversation with the systems engineer Guru Madhavan on this theme: https://evolution-institute.org/ systems-engineering-as-cultural-group-selection-a-conversation -with-guru-madhavan/.
3. https://www.dancingrabbit.org.
4. https://www.ic.org.
5. https://contextualscience.org.
6. In addition to my book *Evolution for Everyone* (Wilson 2007), written for a general audience, a sizable academic literature exists for the value of a TVOL education, including a special issue of the journal *Evolution, Education, and Outreach* (volume 4, issue 1) and an edited volume published by Oxford University Press (Geher et al. 2018). In addition, the EvoS (standing for <u>Evo</u>lutionary <u>Stud</u>-

ies) consortium has its own website and online journal: http://evostudies.org.

7. https://evolution-institute.org.
8. https://www.prosocial.world.
9. https://evolution-institute.org/tvol1000/.
10. https://evolution-institute.org/this-view-of-life/.

Literature Cited

Acemoglu, D., and J. Robinson. 2012. *Why Nations Fail: The Origins of Power, Prosperity, and Poverty*. New York: Crown.

Aktipis, C. A., A. M. Boddy, R. A. Gatenby, J. S. Brown, and C. C. Maley. 2013. "Life History Trade-offs in Cancer Evolution." *Nature Reviews Cancer* 13(12): 883–892. https://doi.org/10.1038/nrc3606.

Aktipis, C. A., V. S. Y. Kwan, K. A. Johnson, S. L. Neuberg, and C. C. Maley. 2011. "Overlooking Evolution: A Systematic Analysis of Cancer Relapse and Therapeutic Resistance Research." *PloS One* 6(11): e26100. https://doi.org/10.1371/journal.pone.0026100.

Aktipis, C. A., and R. M. Nesse. 2013. "Evolutionary Foundations for Cancer Biology." *Evolutionary Applications* 6(1): 144–159. https://doi.org/10.1111/eva.12034.

Amir, A. 2007. *The Jesuit and the Skull: Teilhard de Chardin, Evolution, and the Search for Peking Man*. New York: Riverhead (Penguin).

Ann Arbor Science for the People Collective (ed.). 1977. *Biology as a Social Weapon*. Minneapolis: Burgess.

Barkow, J. H., L. Cosmides, and J. Tooby. 1992. *The Adapted Mind: Evolutionary Psychology and the Generation of Culture*. Oxford: Oxford University Press.

Barr, A. P., ed. 1997. *Thomas Henry Huxley's Place in Science and Letters: Centenary Essays*. Athens: University of Georgia Press.

Barr, R. 2013. "Memory Constraints on Infant Learning from Picture Books, Television, and Touchscreens." *Child Development Perspectives* 7(4): 205–210. https://doi.org/10.1111/cdep.12041.

Barrish, H. H., M. Saunders, and M. M. Wolf. 1969. "Good Behavior Game: Effects of Individual Contingencies for Group Consequences on Disruptive Behavior in a Classroom." *Journal of Applied Behav-*

ior Analysis 2(2): 1311049. www.ncbi.nlm.nih.gov/pmc/articles/PMC1311049/.

Bashford, A., and J. E. Chaplin. 2016. *The New Worlds of Thomas Robert Malthus: Rereading the Principle of Population*. Princeton, NJ: Princeton University Press.

Bateson, P., and K. N. Laland. 2013. "Tinbergen's Four Questions: An Appreciation and an Update." *Trends in Ecology & Evolution* 28: 712–718.

Beck, D. E., and C. C. Cowan. 2006. *Spiral Dynamics: Mastering Values, Leadership and Change: Exploring the New Science of Memetics*. Oxford: Blackwell.

Beckes, L., and J. A. Coan. 2011. "Social Baseline Theory: The Role of Social Proximity in Emotion and Economy of Action." *Social and Personality Psychology Compass* 5(12): 976–988. https://doi.org/10.1111/j.1751-9004.2011.00400.x.

Beckes, L., J. A. Coan, and K. Hasselmo. 2013. "Familiarity Promotes the Blurring of Self and Other in the Neural Representation of Threat." *Social Cognitive and Affective Neuroscience* 8(6), 670–677. https://doi.org/10.1093/scan/nss046.

Beinhocker, E. D. 2006. *Origin of Wealth: Evolution, Complexity, and the Radical Remaking of Economics*. Cambridge, MA: Harvard Business School Press.

Benard, M. F. 2006. "Survival Trade-Offs Between Two Predator-Induced Phenotypes in Pacific Treefrogs (Pseudacris Regilla)." *Ecology* 87(2): 340–346. https://doi.org/10.1890/05-0381.

Bhalla, M., and D. R. Proffitt. 1999. "Visual-Motor Recalibration in Geographical Slant Perception." *Journal of Experimental Psychology: Human Perception and Performance* 25(4): 1076–1096. https://doi.org/10.1037/0096-1523.25.4.1076.

Biglan, A. 2015. *The Nurture Effect: How the Science of Human Behavior Can Improve Our Lives and Our World*. Oakland, CA: New Harbinger Publications.

Bingham, P. M. 1999. "Human Uniqueness: A General Theory." *Quarterly Review of Biology* 74: 133–169.

Bingham, P. M., and J. Souza. 2009. *Death from a Distance and the Birth of a Humane Universe*. BookSurge. Retrieved from http://www.deathfromadistance.com/.

Bjork, D. W. 1993. *B. F. Skinner: A Life*. New York: Basic Books.

Bjorklund, D. F., and B. J. Ellis. 2014. "Children, Childhood, and Development in Evolutionary Perspective." *Developmental Review* 34(3): 225–264. https://doi.org/10.1016/j.dr.2014.05.005.

Bloomfield, S. F., G. A. W. Rook, E. A. Scott, F. Shanahan, R. Stanwell-Smith, and P. Turner. 2016. "Time to Abandon the Hygiene Hypothesis: New Perspectives on Allergic Disease, the Human Microbiome, Infectious Disease Prevention and the Role of Targeted Hygiene." *Perspectives in Public Health* 136(4): 213–224. https://doi.org/10.1177/1757913916650225.

Bodkin, B. B. 1990. *Discordant Harmonies: A New Ecology for the Twenty-First Century*. New York: Oxford University Press.

Boehm, C. 1999. *Hierarchy in the Forest*. Cambridge, MA: Harvard University Press.

———. 2011. *Moral Origins: The Evolution of Virtue, Altruism, and Shame*. New York: Basic Books.

Bowlby, J. 1990. *A Secure Base: Parent-Child Attachment and Healthy Human Development*. New York: Basic Books.

Brookes, M. 2004. *Extreme Measures: The Dark Visions and Bright Ideas of Francis Galton*. London: Bloomsbury Press.

Browne, J. 1995. *Charles Darwin: Voyaging*. New York: Knopf.

———. 2002. *Charles Darwin: The Power of Place*. New York: Knopf.

Burt, A., and R. Trivers. 2006. *Genes in Conflict*. Cambridge, MA: Harvard University Press.

Campbell, D. T. 1974. "Evolutionary Epistemology." In P. A. Schilpp (ed.), *The Philosophy of Karl Popper*, pp. 413–463. LaSalle, IL: Open Court Publishing.

———. 1990. "Levels of Organization, Downward Causation, and the Selection-Theory Approach to Evolutionary Epistemology." In G. Greenberg and E. Tobach (eds.), *Theories of the Evolution of Knowing*, pp. 1–17. Hillsdale, NJ: Lawrence Erlbaum Associates.

———. 1994. "How Individual and Face-to-Face-Group Selection Undermine Firm Selection in Organizational Evolution." In J. A. C. Baum and J. V. Singh (eds.), *Evolutionary Dynamics of Organizations*, pp. 23–38. New York: Oxford University Press.

Chen, X., and T. F. Kelly. 2014. "B-Corps—a Growing Form of Social Enterprise: Tracing Their Progress and Assessing Their Performance." *Journal of Leadership & Organizational Studies* 22(1): 102–114. https://doi.org/10.1177/1548051814532529.

Cho, Y. 1981. *Successful Home Cell Groups*. Alachua, FL: Logos International.

Chudek, M., S. Heller, S. Birch, and J. Henrich. 2012. "Prestige-Biased Cultural Learning: Bystander's Differential Attention to Potential Models Influences Children's Learning." *Evolution and Human Behavior* 33(1): 46–56. https://doi.org/10.1016/j.evolhumbehav.2011.05.005.

Coan, J. A., and J. A. Sbarra. 2015. "Social Baseline Theory: The Social Regulation of Risk and Effort." *Current Opinion in Psychology* 1: 87–91.

Coan, J. A., H. S. Schaefer, and R. J. Davidson. 2006. "Lending a Hand." *Psychological Science* 17(12): 1032–1039. https://doi.org/10.1111/j.1467-9280.2006.01832.x.

Colander, D., and R. Kupers. 2014. *Complexity and the Art of Public Policy: Solving Society's Problems from the Bottom Up*. Princeton, NJ: Princeton University Press.

Cordain, L., S. B. Eaton, J. Brand Miller, S. Lindeberg, and C. Jensen. 2002. "An Evolutionary Analysis of the Aetiology and Pathogenesis of Juvenile-Onset Myopia." *Acta Ophthalmologica Scandinavica* 80(2): 125–135. https://doi.org/10.1034/j.1600-0420.2002.800203.x.

Cox, M., G. Arnold, and S. Villamayor-Tomas. 2010. "A Review of Design Principles for Community-Based Natural Resource Management." *Ecology and Society* 15. Retrieved from http://www.ecologyandsociety.org/vol15/iss4/art38/.

Craig, J. V., and W. M. Muir. 1995. "Group Selection for Adaptation to Multiple-Hen Cages: Beak-Related Mortality, Feathering, and Body Weight Responses." *Poultry Science* 75(3): 294–302.

Crespi, B., and K. Summers. 2005. "Evolutionary Biology of Cancer." *Trends in Ecology & Evolution* 20(10): 545–552. https://doi.org/10.1016/j.tree.2005.07.007.

Csikszentmihalyi, M., K. Rathunde, and S. Whalen. 1993. *Talented Teenagers: The Roots of Success and Failure*. Cambridge, UK: Cambridge University Press.

Dalai Lama, H. H. 2011. *Beyond Religion: Ethics for a Whole World*. New York: Houghton Mifflin Harcourt.

Daly, M., and M. Wilson. 1988. *Homicide*. New York: Aldine de Gruyter.

Darling-Hammond, L., and J. Snyder. 1992. "Curriculum Studies and the Traditions of Inquiry: The Scientific Tradition." In P. W. Jackson (ed.), *Handbook of Research on Curriculum*, pp. 41–78. New York: Macmillan.

Darwin, C. 1859. *The Origin of Species*. London: John Murray.

———. 1871. *The Descent of Man and Selection in Relation to Sex*. 2 vols. London: John Murray.

———. 1887. *The Autobiography of Charles Darwin, 1809–1882. With Original Omissions Restored*. New York: Harcourt Brace.

Davies, N. B., J. R. Krebs, and S. A. West. 2012. *An Introduction to Behavioural Ecology*. 4th ed. Hoboken, NJ: Wiley-Blackwell.

Dawkins, R. 1976. *The Selfish Gene*. Oxford: Oxford University Press.

———. 2006. *The God Delusion*. Boston: Houghton Mifflin.

DeLoache, J. S., C. Chiong, K. Sherman, N. Islam, M. Vanderborght, G. L. Troseth, . . . and K. O'Doherty. 2010. "Do Babies Learn from Baby Media?" *Psychological Science* 21(11): 1570–1574. https://doi.org/10.1177/0956797610384145.

Dethlefsen, L., M. McFall-Ngai, and D. A. Relman. 2007. "An Ecological and Evolutionary Perspective on Human-Microbe Mutualism and Disease." *Nature* 449(7164): 811–818. https://doi.org/10.1038/nature06245.

Dewey, J. (1910). *The Influence of Darwin on Philosophy, and Other Essays on Contemporary Thought*. New York: Henry Holt and Company.

Dobzhansky, T. 1973. "Nothing in Biology Makes Sense Except in the Light of Evolution." *American Biology Teacher* 35: 125–129.

Domitrovich, C. E., C. P. Bradshaw, J. K. Berg, E. T. Pas, K. D. Becker, R. Musci, . . . and N. Ialongo. 2016. "How Do School-Based Prevention Programs Impact Teachers? Findings from a Randomized Trial of an Integrated Classroom Management and Social-Emotional Program." *Prevention Science* 17(3): 325–337. https://doi.org/10.1007/s11121-015-0618-z.

Dugatkin, L. A. 2011. *The Prince of Evolution: Peter Kropotkin's Adventures in Science and Politics*. CreateSpace Independent Publishing Platform.

Ehrenpreis, A., C. Felbinger, and R. Friedmann. 1978. "An Epistle on Brotherly Community as the Highest Command of Love." In

R. Friedmann (ed.), *Brotherly Community: The Highest Command of Love*, pp. 9–77. Rifton, NY: Plough Publishing Co.

Ehrenreich, B., and J. McIntosh. 1997. "The New Creationism: Biology Under Attack." *The Nation*, June 9, 11–16.

Ellis, B., J. Bianchi, V. Griskevicius, and W. Frankenhuis. 2017. "Beyond Risk and Protective Factors: An Adaptation-Based Approach to Resilience." *Perspectives on Psychological Science* 12(4): 561–587.

Ellis, B. J., M. Del Giudice, T. J. Dishion, A. J. Figueredo, P. Gray, V. Griskevicius, . . . and D. S. Wilson. 2012. "The Evolutionary Basis of Risky Adolescent Behavior: Implications for Science, Policy, and Practice." *Developmental Psychology* 48(3): 598–623. https://doi.org/10.1037/a0026220.

Ellis, E. C. 2015. "Ecology in an Anthropogenic Biosphere." *Ecological Monographs* 85(3): 287–331. https://doi.org/10.1890/14-2274.1.

Embry, D. D. 2002. "The Good Behavior Game: A Best Practice Candidate as a Universal Behavioral Vaccine." *Clinical Child and Family Psychology Review* 5: 273–297.

Fessler, D. 2004. "Shame in Two Cultures: Implications for Evolutionary Approaches." *Journal of Cognition and Culture* 4(2): 207–262. https://doi.org/10.1163/1568537041725097.

Flandroy, L., T. Poutahidis, G. Berg, G. Clarke, M. C. Dao, E. Decaestecker, . . . and G. Rook. 2018. "The Impact of Human Activities and Lifestyles on the Interlinked Microbiota and Health of Humans and of Ecosystems." *Science of the Total Environment* 627: 1018–1038. https://doi.org/10.1016/j.scitotenv.2018.01.288.

Folger, R., M. Johnson, and C. Letwin. 2014. "Evolving Concepts of Evolution: The Case of Shame and Guilt." *Social and Personality Psychology Compass* 8(12): 659–671. https://doi.org/10.1111/spc3.12137.

Francis, M. 2007. *Herbert Spencer and the Invention of Modern Life*. Newcastle, UK: Acumen Publishing.

Frank, R. 2011. *The Darwin Economy: Liberty, Competition, and the Common Good*. Princeton, NJ: Princeton University Press.

Fukuyama, F. 2012. *The Origins of Political Order: From Prehuman Times to the French Revolution*. New York: Farrar, Straus & Giroux.

Gallace, A., and C. Spence. 2010. "The Science of Interpersonal Touch: An Overview." *Neuroscience and Biobehavioral Reviews* 34: 246–259.

Gardner, H. 1985. *The Mind's New Science: A History of the Cognitive Revolution*. New York: Basic Books.

Gaynes, R. P., and the American Society for Microbiology. 2011. *Germ Theory: Medical Pioneers in Infectious Diseases*. Washington, DC: ASM Press.

Geary, D. C., and D. B. Berch (eds.). 2016. *Evolutionary Perspectives on Child Development and Education*. Switzerland: Springer International Publishing.

Geher, G., A. C. Gallup, and D. S. Wilson. 2018. *Evolutionary Studies: Darwin's Roadmap to the Curriculum*. Oxford: Oxford University Press.

Gershon, D. 2006. *Low Carbon Diet: A 30-Day Program to Lose 5,000 Pounds*. Berwyn Heights, MD: Empowerment Institute.

———. 2009. *Social Change 2.0: A Blueprint for Reinventing Our World*. White River Junction, VT: High Point.

Gleick, J. 1987. *Chaos: Making a New Science*. New York: Penguin Books.

Goldschmidt, E., and N. Jacobsen. 2014. "Genetic and Environmental Effects on Myopia Development and Progression." *Eye* 28(2): 126–133. https://doi.org/10.1038/eye.2013.254.

Gómez, F., P. López-García, and D. Moreira. 2009. "Molecular Phylogeny of the Ocelloid-Bearing Dinoflagellates *Erythropsidinium* and *Warnowia* (Warnowiaceae, Dinophyceae)." *Journal of Eukaryotic Microbiology* 56(5): 440–445. https://doi.org/10.1111/j.1550-7408.2009.00420.x.

Gould, S. J., and R. C. Lewontin. 1979. "The Spandrels of San Marco and the Panglossian Paradigm: A Critique of the Adaptationist Program." *Proceedings of the Royal Society of London* B205: 581–598.

Grant, A. M. 2013. *Give and Take: A Revolutionary Approach to Success*. New York: Viking.

Gray, P. 2013. *Free to Learn: Why Unleashing the Instinct to Play Will Make Our Children Happier, More Self-Reliant, and Better Students for Life*. New York: Basic Books.

Gregory, S. W., and S. Webster. 1996. "A Nonverbal Signal in Voices of Interview Partners Effectively Predicts Communication Accommodation and Social Status Perceptions." *Journal of Personality and Social Psychology* 70(6): 1231–1240. https://doi.org/10.1037/0022-3514.70.6.1231.

Grunbaum, A. 1984. *The Foundations of Psychoanalysis: A Philosophical Critique*. Berkeley: University of California Press.

Haig, D. 1993. "Genetic Conflicts in Human Pregnancy." *Quarterly Review of Biology* 68(4): 495–532. https://doi.org/10.1086/418300.

———. 2015. "Maternal-Fetal Conflict, Genomic Imprinting and Mammalian Vulnerabilities to Cancer." *Philosophical Transactions of the Royal Society of London B: Biological Sciences* 370(1673). Retrieved from http://rstb.royalsocietypublishing.org/content/370/1673/20140178.

Hanski, I., L. von Hertzen, N. Fyhrquist, K. Koskinen, K. Torppa, T. Laatikainen . . . and T. Haahtela. 2012. "Environmental Biodiversity, Human Microbiota, and Allergy Are Interrelated." *Proceedings of the National Academy of Sciences of the United States of America* 109(21): 8334–8339. https://doi.org/10.1073/pnas.1205624109.

Hardin, G. 1968. "The Tragedy of the Commons." *Science* 162: 1243–1248.

Hartberg, Y. M., and D. S. Wilson. 2016. "Sacred Text as Cultural Genome: An Inheritance Mechanism and Method for Studying Cultural Evolution." *Religion, Brain & Behavior* 7(3): 1–13. https://doi.org/10.1080/2153599X.2016.1195766.

Hayes, S. C., B. T. Sanford, and F. T. Chin. 2017. "Carrying the Baton: Evolution Science and a Contextual Behavioral Analysis of Language and Cognition." *Journal of Contextual Behavioral Science* 6(3): 314–328. https://doi.org/10.1016/j.jcbs.2017.01.002.

Hayes, S. C., and S. Smith. 2005. *Get Out of Your Mind and into Your Life: The New Acceptance and Commitment Therapy*. Oakland, CA: New Harbinger Press.

Hayes, S. C., K. Strosahl, and K. G. Wilson. 2011. *Acceptance and Commitment Therapy: The Process and Practice of Mindful Change*. 2nd ed. New York: Guilford.

Hedges, C. 2002. *War Is a Force That Gives Us Meaning*. PublicAffairs. Retrieved from http://www.amazon.com/Force-That-Gives-Meaning-ebook/dp/B006MK0HYQ.

Henrich, J. 2015. *The Secret of Our Success: How Culture Is Driving Human Evolution, Domesticating Our Species, and Making Us Smarter*. Princeton, NJ: Princeton University Press.

Henrich, J., R. Boyd, and P. J. Richerson. 2008. "Five Misunderstandings About Cultural Evolution." *Human Nature* 19: 119–137.

Higgs, P. G., and N. Lehman. 2015. "The RNA World: Molecular Cooperation at the Origins of Life." *Nature Reviews Genetics* 16(1): 7–17.

Hodgson, G. M. 2004. "Social Darwinism in Anglophone Academic Journals: A Contribution to the History of the Term." *Journal of Historical Sociology* 17(4): 428–463.

———. 2007. "Taxonomizing the Relationship Between Biology and Economics: A Very Long Engagement." *Journal of Bioeconomics* 9: 169–185.

———. 2015. *Conceptualizing Capitalism: Institutions, Evolution, Future.* Chicago: University of Chicago Press.

Hodgson, G. M., and K. Thorbjorn. 2010. *Darwin's Conjecture: The Search for General Principles of Social and Economic Evolution.* Chicago: University of Chicago Press.

Hofstadter, R. 1959. *Social Darwinism in American Thought.* Boston: Beacon Press.

Holldobler, B., and E. O. Wilson. 2008. *The Superorganisms.* New York: Norton.

Horne, S. D., S. A. Pollick, and H. H. Q. Heng. 2015. "Evolutionary Mechanism Unifies the Hallmarks of Cancer." *International Journal of Cancer. Journal International Du Cancer* 136(9): 2012–2021. https://doi.org/10.1002/ijc.29031.

Hrdy, S. 2011. *Mothers and Others: The Evolutionary Origins of Human Understanding.* Cambridge, MA: Belknap.

Hull, D. L. 1990. *Science as a Process: An Evolutionary Account of the Social and Conceptual Development of Science.* Chicago: University of Chicago Press.

Hwang, V., and G. Horowitt. 2012. *The Rainforest: The Secret to Building the Next Silicon Valley.* Los Altos Hills, CA: Regenwald.

Jablonka, E., and M. Lamb. 2006. *Evolution in Four Dimensions: Genetic, Epigenetic, Behavioral, and Symbolic Variation in the History of Life.* Cambridge, MA: MIT Press.

Jackall, R. 2009. *Moral Mazes: The World of Corporate Managers.* New York: Oxford University Press.

Jacobs, J. 1961. *The Death and Life of Great American Cities.* New York: Vintage.

Jansen, G., R. Gatenby, and C. A. Aktipis. 2015. "Opinion: Control vs.

Eradication: Applying Infectious Disease Treatment Strategies to Cancer: Fig. 1." *Proceedings of the National Academy of Sciences* 112(4): 937–938. https://doi.org/10.1073/pnas.1420297111.

Jeffcoat, T., and S. C. Hayes. 2012. "A Randomized Trial of ACT Bibliotherapy on the Mental Health of K-12 Teachers and Staff." *Behaviour Research and Therapy* 50(9): 571–579. https://doi.org/10.1016/j.brat.2012.05.008.

Johnson, K., and S. L. Coates. 2001. *Nabokov's Blues: The Scientific Odyssey of a Literary Genius*. New York: McGraw-Hill.

Johnson, K., and D. Ord. 2013. *The Coming Interspiritual Age*. Vancouver: Namaste.

Jones, D. S. 2012. *Masters of the Universe: Hayek, Friedman, and the Birth of Neoliberal Politics*. Princeton, NJ: Princeton University Press. Retrieved from http://www.amazon.com/Masters-Universe-Friedman-Neoliberal-Politics/dp/0691151571/ref=pd_sim_b_1?ie=UTF8&refRID=1ZMKMGNWTPZPQ2FBQHXV.

Kellam, S. G., A. C. L. Mackenzie, C. H. Brown, J. M. Poduska, W. Wang, H. Petras, and H. C. Wilcox. 2011. "The Good Behavior Game and the Future of Prevention and Treatment." *Addiction Science & Clinical Practice* 6(1): 73–84. Retrieved from http://www.scopus.com/inward/record.url?eid=2-s2.0–84864950294&partnerID ZOtx3y1.

Kellam, S. G., W. Wang, A. C. L. Mackenzie, C. H. Brown, D. C. Ompad, F. Or, . . . and A. Windham. 2014. "The Impact of the Good Behavior Game, a Universal Classroom-Based Preventive Intervention in First and Second Grades, on High-Risk Sexual Behaviors and Drug Abuse and Dependence Disorders into Young Adulthood." *Prevention Science* 15(S1): 6–18. https://doi.org/10.1007/s11121-012-0296-z.

King, U. 2015. *Spirit of Fire: The Life and Vision of Teilhard de Chardin*. Maryknoll, NY: Orbis Books.

Kricher, J. 2009. *The Balance of Nature: Ecology's Enduring Myth*. Princeton, NJ: Princeton University Press.

Laland, K. N. 2017. *Darwin's Unfinished Symphony: How Culture Made the Human Mind*. Princeton, NJ: Princeton University Press.

Lickliter, R., and P. Virkar. 1989. "Intersensory Functioning in Bobwhite Quail Chicks: Early Sensory Dominance." *Developmental Psychobiology* 22(7): 651–667. https://doi.org/10.1002/dev.420220702.

Lloyd, L., D. S. Wilson, and E. Sober. 2014. "Evolutionary Mismatch and What to Do About It: A Basic Tutorial." Evolution Institute White Paper. Retrieved from http://evolution-institute.org/sites/default/files/articles/Mismatch-Sept-24–2011.pdf.

Mandeville, B. (1714) 1988. *The Fable of the Bees: or Private Vices, Publick Benefits.* Indianapolis, IN: Liberty Fund.

Margulis, L. 1970. *Origin of Eukaryotic Cells.* New Haven, CT: Yale University Press.

Martincorena, I., A. Roshan, M. Gerstung, P. Ellis, P. Van Loo, S. Mclaren, . . . and P. J. Campbell. 2015. "High Burden and Pervasive Positive Selection of Somatic Mutations in Normal Human Skin." *Science* 348(6237): 880–887.

Matta, N. F., and R. N. Ashkenas. 2003. "Why Good Projects Fail Anyway." *Harvard Business Review* 81(9):109–114, 134.

Matta, N., and P. Morgan. 2011. "Local Empowerment Through Rapid Results." *Stanford Social Innovation Review* (Summer) 201: 49–55.

Maynard Smith, J., and E. Szathmáry. 1995. *The Major Transitions in Evolution.* New York: W. H. Freeman.

———. 1999. *The Origins of Life: From the Birth of Life to the Origin of Language.* Oxford: Oxford University Press.

McCauley, R. N. 2011. *Why Religion Is Natural and Science Is Not.* New York: Oxford University Press.

Menand, L. 2001. *The Metaphysical Club: A Story of Ideas in America.* New York: Farrar, Straus & Giroux.

Miller, A. H., and C. L. Raison. 2016. "The Role of Inflammation in Depression: From Evolutionary Imperative to Modern Treatment Target." *Nature Reviews Immunology* 16(1): 22–34. https://doi.org/10.1038/nri.2015.5.

Miller, G. A. 2003. "The Cognitive Revolution: A Historical Perspective." *Trends in Cognitive Sciences* 7(3): 141–144. https://doi.org/10.1016/S1364-6613(03)00029-9.

Morgan, R. W., J. S. Speakman, and S. E. Grimshaw. 1975. "Inuit Myopia: An Environmentally Induced 'Epidemic'?" *Canadian Medical Association Journal* 112(5): 575–577.

Muir, W. M. 1995. "Group Selection for Adaptation to Multiple-Hen Cages: Selection Program and Direct Responses." *Poultry Science* 75(4): 447–458.

Muir, W. M., M. J. Wade, P. Bijma, and E. D. Ester. 2010. "Group Selection and Social Evolution in Domesticated Chickens." *Evolutionary Applications* 3: 453–465.

Muto, T., S. C. Hayes, and T. Jeffcoat. 2011. "The Effectiveness of Acceptance and Commitment Therapy Bibliotherapy for Enhancing the Psychological Health of Japanese College Students Living Abroad." *Behavior Therapy* 42(2): 323–335. https://doi.org/10.1016/j .beth.2010.08.009.

Nabokov, V. V., S. H. Blackwell, H. Stephen, and K. Johnson. 2016. *Fine Lines: Vladimir Nabokov's Scientific Art*. New Haven, CT: Yale University Press.

Naess, A., A. R. Drengson, and B. Devall. 2010. *Ecology of Wisdom: Writings by Arne Naess*. Berkeley, CA: Counterpoint.

Nelson, R. R., and S. G. Winter. 1982. *An Evolutionary Theory of Economic Change*. Cambridge, MA: Harvard University Press.

Nichols, R. 2015. "Civilizing Humans with Shame: How Early Confucians Altered Inherited Evolutionary Norms Through Cultural Programming to Increase Social Harmony." *Journal of Cognition and Culture* 15(3–4): 254–284. https://doi.org/10.1163/15685373 -12342150.

Novoa, A. 2016. "Social Darwinism: A Case of Designed Ventriloquism." In D. S. Wilson and E. M. Johnson (eds.), *Truth and Reconciliation for Social Darwinism*. Special issue of *This View of Life*. https://evolution -institute.org/wp-content/uploads/2016/11/2Social-Darwinism_ Publication.pdf.

Nuland, S. B. 2003. *The Doctors' Plague: Germs, Childbed Fever, and the Strange Story of Ignác Semmelweis*. New York: W. W. Norton.

Nuttall, G. 2012. *Sharing Success: The Nuttall Review of Employee Ownership*. https://www.gov.uk/government/uploads/system/uploads/ attachment_data/file/31706/12-933-sharing-success-nuttall-review -employee-ownership.pdf.

Oakerson, R. J., and J. D. W. Clifton. 2011. "Neighborhood Decline as a Tragedy of the Commons: Conditions of Neighborhood Turnaround on Buffalo's West Side." *Workshop in Political Theory and Policy Analysis* W11–26.

O'Brien, D. T. 2018. *The Urban Commons: Leveraging Digital Data and*

Technology to Better Understand and Manage the Maintenance of City Neighborhoods. Cambridge, MA: Harvard University Press.

O'Brien, D. T., A. C. Gallup, and D. S. Wilson. 2012. "Residential Mobility and Prosocial Development Within a Single City." *American Journal of Community Psychology* 50(1–2): 26–36. https://doi.org/10.1007/s10464-011-9468-4.

O'Brien, D. T., and D. S. Wilson. 2011. Community Perception: The Ability to Assess the Safety of Unfamiliar Neighborhoods and Respond Adaptively. *Journal of Personality and Social Psychology* 100(4): 606–620. https://doi.org/10.1037/a0022803.

O'Brien, D. T., D. S. Wilson, and P. H. Hawley. 2009. "'Evolution for Everyone': A Course That Expands Evolutionary Theory Beyond the Biological Sciences." *Evolution: Education and Outreach* 2(3): 445–457. https://doi.org/10.1007/s12052-009-0161-0.

O'Connell, M. E., T. Boat, and K. E. Warner (eds.). 2009. *Preventing Mental, Emotional, and Behavioral Disorders Among Young People: Progress and Possibilities*. Washington, DC: National Academies Press.

O'Donohue, W. T., D. Henderson, S. C. Hayes, J. Fisher, and L. Hayes. 2001. *A History of the Behavioral Therapies: Founders' Personal Histories*. Oakland, CA: Context Press.

Ofek, H. 2001. *Second Nature: Economics Origins of Human Evolution*. Cambridge: Cambridge University Press.

Ostrom, E. 1990. *Governing the Commons: The Evolution of Institutions for Collective Action*. Cambridge: Cambridge University Press.

———. 2010(a). "Polycentric Systems for Coping with Collective Action and Global Environmental Change." *Global Environmental Change* 20: 550–557.

———. 2010(b). "Beyond Markets and States: Polycentric Governance of Complex Economic Systems." *American Economic Review* 100: 1–33.

———. 2013. "Do Institutions for Collective Action Evolve?" *Journal of Bioeconomics* 16(1): 3–30. https://doi.org/10.1007/s10818-013-9154-8.

Paradis, J. G., and G. C. Williams. 2016. *Evolution and Ethics: T. H. Huxley's Evolution and Ethics with New Essays on Its Victorian and Sociobiological Context*. Princeton, NJ: Princeton University Press.

Paul, R. A. 2015. *Mixed Messages: Cultural and Genetic Inheritance in the Constitution of Human Society*. Chicago: University of Chicago Press.

Pennebaker, J. W. 2010. *Writing to Heal: A Guided Journal for Recovering from Trauma and Emotional Upheaval*. Oakland, CA: New Harbinger.

Pennebaker, J. W., and J. D. Seagal. 1999. "Forming a Story: The Health Benefits of Narrative." *Journal of Clinical Psychology* 55: 1243–1254.

Pfeffer, J. 1998. *The Human Equation: Building Profits by Putting People First*. Cambridge, MA: Harvard Business Review Press.

Pickett, K., and J. B. Wilkinson. 2009. *The Spirit Level: Why Greater Equality Makes Societies Stronger*. London: Bloomsbury Press.

Polk, K. L., B. Schoendorff, and K. G. Wilson. 2014. *The ACT Matrix: A New Approach to Building Psychological Flexibility Across Settings and Populations*. Oakland, CA: Context Press.

Prinz, R. J., M. R. Sanders, C. J. Shapiro, D. J. Whitaker, and J. R. Lutzker. 2009. "Population-Based Prevention of Child Maltreatment: The U.S. Triple P Population Trial." *Prevention Science* 10: 1–12.

Provine, W. B. 1986. *Sewall Wright and Evolutionary Biology*. Chicago: University of Chicago Press.

———. 2001. *The Origins of Theoretical Population Genetics*. Chicago: University of Chicago Press.

Putnam, R. D. 2000. *Bowling Alone: The Collapse and Revival of American Community*. New York: Simon & Schuster.

Radesky, J. S., M. Silverstein, B. Zuckerman, and D. A. Christakis. 2014. "Infant Self-Regulation and Early Childhood Media Exposure." *Pediatrics* 133(5). Retrieved from http://pediatrics.aappublications.org/content/133/5/e1172.short.

Richards, R. J. 2013. *Was Hitler a Darwinian? Disputed Questions in the History of Evolutionary Theory*. Chicago: University of Chicago Press.

Richerson, P. J., and R. Boyd. 2005. *Not by Genes Alone: How Culture Transformed Human Evolution*. Chicago: University of Chicago Press.

Rook, G. A. W. 2012. "Hygiene Hypothesis and Autoimmune Diseases." *Clinical Reviews in Allergy & Immunology* 42(1): 5–15. https://doi.org/10.1007/s12016-011-8285-8.

———. 2013. "Regulation of the Immune System by Biodiversity from the Natural Environment: An Ecosystem Service Essential to Health." *Proceedings of the National Academy of Sciences of the United*

States of America 110(46): 18360–18367. https://doi.org/10.1073/pnas.1313731110.

Rook, G. A. W., and C. A. Lowry. 2008. "The Hygiene Hypothesis and Psychiatric Disorders." *Trends in Immunology* 29(4): 150–158. https://doi.org/10.1016/J.IT.2008.01.002.

Rother, M. 2009. *Toyota Kata: Managing People for Improvement, Adaptiveness, and Superior Results*. New York: McGraw-Hill.

———. 2017. *The Toyota Kata Practice Guide*. New York: McGraw-Hill.

Rother, M., G. Aulinger, and L. Wagner. 2017. *Toyota Kata Culture: Building Organizational Capability and Mindset Through Kata Coaching*. New York: McGraw-Hill.

Sampson, R. J. 2003. "The Neighborhood Context of Well-Being." *Perspectives in Biology and Medicine* 46: S53–S64.

Schaffer, R. H., and R. Ashkenas. 2007. *Rapid Results! How 100-Day Projects Build the Capacity for Large-Scale Change*. New York: Jossey-Bass.

Schwab, I. R. 2012. *Evolution's Witness: How Eyes Evolved*. New York: Oxford University Press.

Schweinhart, L. J., and D. P. Weikart. 1997. "The High/Scope Pre-school Curriculum Comparison Study Through Age 23." *Early Childhood Research and Practice* 12: 117–143.

Seeley, T. 1996. *The Wisdom of the Hive*. Cambridge, MA: Harvard University Press.

———. 2010. *Honeybee Democracy*. Princeton, NJ: Princeton University Press.

Sender, R., S. Fuchs, and R. Milo. 2016. "Revised Estimates for the Number of Human and Bacteria Cells in the Body." *PLoS Biology* 14(8): 1–14. https://doi.org/10.1371/journal.pbio.1002533.

Senor, D., and S. Singer. 2009. *Start-up Nation: The Story of Israel's Economic Miracle*. New York: Twelve Books.

Shapin, S. 1995. *A Social History of Truth: Civility and Science in Seventeenth-Century England*. Chicago: University of Chicago Press.

Sherwin, J. C., M. H. Reacher, R. H. Keogh, A. P. Khawaja, D. A. Mackey, and P. J. Foster. 2012. "The Association Between Time Spent Outdoors and Myopia in Children and Adolescents: A Systematic Review and Meta-Analysis." *Ophthalmology* 119(10): 2141–2151. https://doi.org/10.1016/j.ophtha.2012.04.020.

Skinner, B. F. 1981. "Selection by Consequences." *Science* 213: 501–504.

Smail, D. L. 2008. *On Deep History and the Brain*. Berkeley, CA: University of California Press.

Smit, Y., M. J. H. Huibers, J. P. A. Ioannidis, R. van Dyck, W. van Tilburg, and A. Arntz. 2012. "The Effectiveness of Long-Term Psychoanalytic Psychotherapy—a Meta-Analysis of Randomized Controlled Trials." *Clinical Psychology Review* 32(2): 81–92. https://doi.org/10.1016/J.CPR.2011.11.003.

Solnit, R. 2009. *A Paradise Built in Hell: The Extraordinary Communities That Arise in Disasters*. New York: Viking.

Sompayrac, L. M. 2008. *How the Immune System Works*. 3rd ed. Hoboken, NJ: Wiley, Blackwell.

Sosis, R. 2000. "Religion and Intragroup Cooperation: Preliminary Results of a Comparative Analysis of Utopian Communities." *Cross-Cultural Research* 34: 70–87.

Sosis, R., U. Schjoedt, J. Bulbulia, and W. J. Wildman. 2017. "Wilson's 15-Year-Old Cathedral." *Religion, Brain & Behavior* 7(2): 95–97. https://doi.org/10.1080/2153599X.2017.1314409.

Svensson, E. I., and R. Calsbeek. 2012. *The Adaptive Landscape in Evolutionary Biology*. Oxford: Oxford University Press.

Swaab, R. I., M. Schaerer, E. M. Anicich, R. Ronay, and A. D. Galinsky. 2014. "The Too-Much-Talent Effect: Team Interdependence Determines When More Talent Is Too Much or Not Enough." *Psychological Science* 25(8): 1581–1591. https://doi.org/10.1177/0956797614537280.

Szathmáry, E., and J. Maynard Smith. 1997. "From Replicators to Reproducers: The First Major Transitions Leading to Life." *Journal of Theoretical Biology* 187(4): 555–571. https://doi.org/10.1006/jtbi.1996.0389.

Sznycer, D., J. Tooby, L. Cosmides, R. Porat, S. Shalvi, and E. Halperin. 2016. "Shame Closely Tracks the Threat of Devaluation by Others, Even Across Cultures." *Proceedings of the National Academy of Sciences of the United States of America* 113(10): 2625–2630. https://doi.org/10.1073/pnas.1514699113.

Tanaka, H., A. Yagi, A. Komiya, N. Mifune, and Y. Ohtsubo. 2015. "Shame-Prone People Are More Likely to Punish Themselves: A Test of the Reputation-Maintenance Explanation for Self-Punishment."

Evolutionary Behavioral Sciences 9(1): 1–7. https://doi.org/10.1037/ebs0000016.

Teilhard de Chardin, P. 1959. *The Phenomenon of Man.* New York: Collins.

Tinbergen, N. 1963. "On Aims and Methods of Ethology." *Zeitschrift Für Tierpsychologie* 20: 410–433.

———. 1969. *The Curious Naturalist.* New York: Anchor Books.

Trivers, R. 1972. "Parental Investment and Sexual Selection." In B. Campbell (ed.), *Sexual Selection and the Descent of Man.* Chicago: Aldine.

Turchin, P. 2005. *War and Peace and War.* Upper Saddle River, NJ: Pi Press.

———. 2010. "Warfare and the Evolution of Social Complexity: A Multilevel Selection Approach." *Structure and Dynamics* 4(3), 1–37.

———. 2015. *Ultrasociety: How 10,000 Years of War Made Humans the Greatest Cooperators on Earth.* Storrs, CT: Beresta Books.

———. 2016. *Ages of Discord: A Structural-Demographic Analysis of American History.* Storrs, CT: Beresta Books.

Vasas, V., C. Fernando, M. Santos, S. Kauffman, and E. Szathmáry. 2011. "Evolution Before Genes." *Biology Direct* 7(1). https://doi.org/10.1186/1745-6150-7-1.

Veblen, T. 1898. "Why Is Economics Not an Evolutionary Science?" *Quarterly Journal of Economics* 12: 373–397.

Wagner, G. P. 1996. "Perspective: Complex Adaptations and the Evolution of Evolvability." *Evolution* 50: 967–976.

Walton, G. M., and G. L. Cohen. 2011. "A Brief Social-Belonging Intervention Improves Academic and Health Outcomes of Minority Students." *Science* 331(6023): 1447–1451. https://doi.org/10.1126/science.1198364.

Welbourne, T., and A. Andrews. 1996. "Predicting Performance of Initial Public Offering Firms: Should HRM Be in the Equation?" *Academy of Management Journal* 39: 891–919.

Wilson, D. S. 1973. "Food Size Selection Among Copepods." *Ecology* 54: 909–914.

———. 1988. "Holism and Reductionism in Evolutionary Biology." *Oikos* 53: 269–273.

———. 2002. *Darwin's Cathedral: Evolution, Religion and the Nature of Society*. Chicago: University of Chicago Press.

———. 2005(a). "Evolutionary Social Constructivism." In J. Gottshcall and D. S. Wilson (eds.), *The Literary Animal: Evolution and the Nature of Narrative*, vol. 2005, pp. 20–37. Evanston, IL: Northwestern University Press.

———. 2005(b). "Evolution for Everyone: How to Increase Acceptance of, Interest in, and Knowledge About Evolution." *Public Library of Science (PLoS) Biology* 3: 1001–1008.

———. 2007. *Evolution for Everyone: How Darwin's Theory Can Change the Way We Think About Our Lives*. New York: Delacorte.

———. 2011. *The Neighborhood Project: Using Evolution to Improve My City, One Block at a Time*. New York: Little, Brown.

———. 2015. *Does Altruism Exist? Culture, Genes, and the Welfare of Others*. New Haven, CT: Yale University Press.

Wilson, D. S., and J. M. Gowdy. 2014. "Human Ultrasociality and the Invisible Hand: Foundational Developments in Evolutionary Science Alter a Foundational Concept in Economics." *Journal of Bioeconomics* 17(1): 37–52. https://doi.org/10.1007/s10818-014-9192-x.

Wilson, D. S., Y. Hartberg, I. MacDonald, J. A. Lanman, and H. Whitehouse. 2017. "The Nature of Religious Diversity: A Cultural Ecosystem Approach." *Religion, Brain & Behavior* 7(2): 134–153. https://doi.org/10.1080/2153599X.2015.1132243.

Wilson, D. S., and S. C. Hayes. 2018. *Evolution and Contextual Behavioral Science: An Integrated Framework for Understanding, Predicting, and Influencing Behavior*. Menlo Park, CA: New Harbinger Press.

Wilson, D. S., S. C. Hayes, A. Biglan, and D. Embry. 2014. "Evolving the Future: Toward a Science of Intentional Change." *Behavioral and Brain Sciences* 37: 395–460. Retrieved from http://journals.cambridge.org/repo_A93SJz6p.

Wilson, D. S., and E. M. Johnson. 2016a. "Truth and Reconciliation for Social Darwinism." In D. S. Wilson and E. M. Johnson (eds.), *Truth and Reconciliation for Social Darwinism*. Special issue of *This View of Life*. https://evolution-institute.org/wp-content/uploads/2016/11/2Social-Darwinism_Publication.pdf.

Wilson, D. S., and E. M. Johnson (eds.). 2016b. *Truth and Reconciliation for Social Darwinism*. Special issue of *This View of Life*. https://

evolution-institute.org/wp-content/uploads/2016/11/2Social -Darwinism_Publication.pdf.

Wilson, D. S., R. A. Kauffman, and M. S. Purdy. 2011. "A Program for At-Risk High School Students Informed by Evolutionary Science." *PLoS ONE* 6(11): e27826. https://doi.org/10.1371/journal .pone.0027826.

Wilson, D. S., T. F. Kelly, M. M. Philip, and X. Chen. 2015. *Doing Well by Doing Good: An Evolution Institute Report on Socially Responsible Businesses*. Evolution Institute Report: https://evolution-institute.org/ wp-content/uploads/2016/01/EI-Report-Doing-Well-By-Doing -Good.pdf.

Wilson, D. S., and A. Kirman. 2016. *Complexity and Evolution: Toward a New Synthesis for Economics*. Cambridge, MA: MIT Press.

Wilson, D. S., D. T. O'Brien, and A. Sesma. 2009. "Human Prosociality from an Evolutionary Perspective: Variation and Correlations on a City-wide Scale." *Evolution and Human Behavior* 30: 190–200.

Wilson, D. S., E. Ostrom, and M. E. Cox. 2013. "Generalizing the Core Design Principles for the Efficacy of Groups." *Journal of Economic Behavior & Organization* 90: S21–S32. https://doi.org/10.1016/j .jebo.2012.12.010.

Wilson, D. S., and E. O. Wilson. 2007. "Rethinking the Theoretical Foundation of Sociobiology." *Quarterly Review of Biology* 82: 327– 348.

Wilson, E. O. 1975. *Sociobiology: The New Synthesis*. Cambridge, MA: Harvard University Press.

Wilson, T. D. 2015. *Redirect: The Surprising New Science of Psychological Change*. New York: Back Bay Books.

Wright, J. B. 2005. "Adam Smith and Greed." *Journal of Private Enterprise* 21: 46–58. Retrieved from http://journal.apee.org/index.php/ Fall2005_4.

Yong, E. 2006. *I Contain Multitudes: The Microbes Within Us and a Grander View of Life*. New York: Ecco.

Zimmerman, F. J., D. A. Christakis, and A. N. Meltzoff. 2007. "Associations Between Media Viewing and Language Development in Children Under Age 2 Years." *Journal of Pediatrics* 151(4): 364–368. https://doi.org/10.1016/j.jpeds.2007.04.071.

Index

Page numbers in *italics* refer to illustrations.

ILLUSTRATION CREDITS

Page   xii Sokoljan/Wikimedia Commons, under Creative Commons 3.0

Page   9 Science History Images/Alamy Stock Photo

Page   17 Cartoon by Jim Morin, copyright © 2012. First published in *The Miami Herald* on April 13, 2012. Used by permission of Jim Morin.

Page   18 Courtesy of Geoffrey M. Hodgson

Page   37 Jan Arkesteijn/Rob Mieremet (ANEFO)/Wikimedia Commons, under Creative Commons 3.0

Page   38 Illustration by Jennifer Campbell-Smith

Page   39 Kichigin/Shutterstock

Page   44 Loulouka1/Shutterstock

Page   46 Photograph by Greg L. Kohuth, courtesy of Michigan State University

Page   53 Courtesy of Antony J. Durston

Page   56 Illustration by Jennifer Campbell-Smith

Page   61 Kateryna Kon/Shutterstock

Page   66 メルビル/Wikimedia Commons, under Creative Commons 4.0

Page   68 Paul Inkles/flickr, under Creative Commons 2.0

Page   81 Courtesy of Inigo Martincorena, Wellcome Sanger Institute, Genome Research Limited

Page   85 Courtesy of William Muir

Page   86 Courtesy of William Muir

Page   95 National Institute of Allergy and Infectious Diseases

Page   98 By permission of Michael Benard

Page 115 Courtesy of the Ostrom Workshop/Indiana University

Page 119 Jorge Moro/Shutterstock

Page 124 Creativa Images/Shutterstock

Page 128 Courtesy of Carolyn Wilczynski

Page 130 Courtesy of Dennis Embry

Page 152 logoboom/Shutterstock

Page 158 Erika Cross/Shutterstock
Page 169 Courtesy of Alan Honick
Page 186 Courtesy of Dr. Kate E. Pickett and Dr. Gerald S. Wilkinson
Page 189 Courtesy of Peter Turchin
Page 201 Sergey Merzliakov/Shutterstock